AUSTRALIAN GREEN HOME & GARDEN

Robin Stewart grew up in Melbourne where she qualified as a primary school teacher before moving to country Victoria.

With her husband Doug, she combined stud sheep and cattle breeding with a teaching career and caring for orphaned wildlife, moving to King Island in the 1980s. Here they owned a sheep property, bred Irish Setters and established a penguin banding program.

Robin now lives on Phillip Island with her husband, a Great Dane, an Irish Setter, a Border Collie and two Siamese cats. She has been writing full-time for the past nine years.

Other books by Robin Stewart include:

Robin Stewart's Chemical Free Home

Robin Stewart's Chemical Free Pest Control

The Clean House Effect

From Seeds to Leaves: A Complete Guide to Growing Australian Trees and Shrubs from Seed

New Faces: The Complete Book of Alternative Pets (CBC Book of the Year 1995)

Envirocat: A New Approach to Caring for Your Cat and Protecting Native Wildlife

The Dog Book: How to Choose a Dog That Suits Your Personality and Lifestyle

Alternative Pets: From Budgies and Yabbies to Rabbits and Rats

Wombat and Koala: Bush Babies Solo Series

Moonbird, a novel for teenagers

AUSTRALIAN GREEN HOME & GARDEN

Practical
and inexpensive
ways to reduce the
use of chemicals
in your home
and garden

Robin Stewart

For people who recognise that we're part of the environment, rather than separate from it.

Produced by Black Inc.

Level 5, 289 Flinders Lane
Melbourne VIC 3000

Reprinted 2006

National Library of Australia Cataloguing-in-publication data:

Stewart, Robin E. (Robin Elaine), 1943- .

Australian green home and garden.

ISBN 1 86395 323 X.

1. Organic gardening. I. Title.

635.0484

Publisher's Note

The hints and remedies contained within these pages are suggestions only.

It is always wise to experiment, prior to application on a larger scale, due to the wide variation of materials used in the manufacturing process.

Every effort has been made to ensure that this book is free from error or omissions. However, the publisher, the author, the editor, or their respective employees or agents, shall not accept responsibility for injury, loss, or damage occasioned to any person acting or refraining from action as a result of material in this book whether or not such injury, loss, damage is in any way due to any negligent act or omission, breach of duty or default on the part of the publisher, the author, the editor or their respective employees or agents.

Contents

Preface

Living green is about optimism.

Optimism for a truly sustainable future.

Sustainable not only for our species, but for all other forms of life.

For the long-term good of planet Earth, and not exclusively for the human race at this point in time.

This book suggests how you, as an individual, can attune your everyday living to the clean and the green.

How you can join the growing green movement and live in harmony with the environment.

Feel part of the environment, rather than separate from it.

Optimistic that we have a future on this planet.

Introducing the Green Concept

In many ways I've been researching this book all my life. For as long as I can remember I've had a passion for all living things, along with a deep respect for the environment. Being green is my natural state.

Over the years I've watched with dismay as people have become more and more isolated from their environment. Times are changing, however, and these same people are now beginning to crave a more natural lifestyle; are becoming more environmentally aware; are starting to think green.

Now, in the twenty-first century, we need to combine the wisdom of the past with the needs and advances of today. After all, tribal cultures learned, long ago, that if you don't respect your environment, you die. It's as simple as that.

Green living begins at home, using products that are both people-friendly and environmentally safe. We need to move forward, beyond the use of pesticides and chlorine, ammonia and phosphate-based cleaning products. There ARE green alternatives, solutions to everyday household tasks, that are:

- Effective.
- Quick and easy to apply.
- Inexpensive.
- And above all else, healthy for people and healthy for the environment.

People cannot be separated from the environment, nor the environment separated from people. We are one and the same! The way we relate to our environment (including all forms of life) tells us a lot about ourselves. If we respond with understanding, respect and sensitivity, we are likely to show the same understanding, respect and sensitivity towards other people – and towards ourselves.

In modern times, it's a challenge to live green; to live in harmony with nature in its entirety. In an attempt to recreate our natural environment – to be green – we nurture our gardens, create water features and garden ponds. We include dogs, cats and all manner of other pets within our family circle.

Indoors we strive to do likewise, by grasping little gems of nature in the form of paintings and photos of inspirational scenery, aquariums of exotic fish, flower arrangements and indoor plants. Deep down, we know that these things create feelings of calm tranquillity – let us feel whole again. Deep down we know that we're part of the environment and yearn to be closer.

By following the suggestions in this book you will find that living green improves the quality of your life – restores balance and a sense of connection.

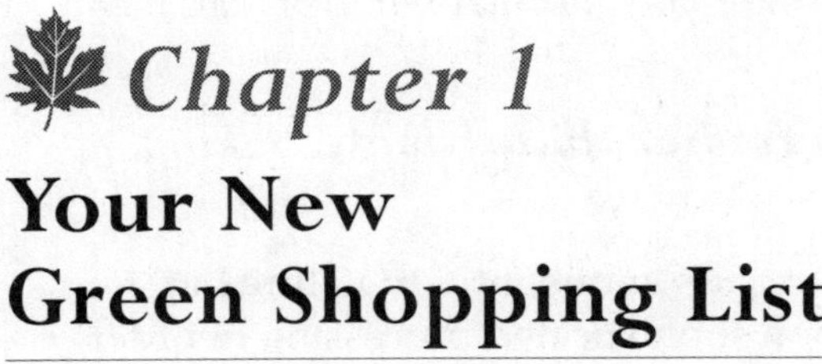

Chapter 1

Your New Green Shopping List

'Let the buyer be informed' is our new catch-cry, rather than 'Let the buyer beware.' We must always remember that as consumers we have power if we choose to use it. But we need to be both aware and informed.

TV advertising doesn't have to influence our thinking, despite the constant brain-washing that tells us we need to buy a different commercial cleaning product for every surface, appliance and room of the house. It's all too common to have up to 20 different cleaning products beneath the kitchen sink.

People's attitudes and shopping patterns are changing, however. Clean and green is now recognised as a modern concept, for modern living. Generally speaking, if you can safely eat a product, that product is also environmentally kind as a cleaner.

When planning your change-over to green household cleaners, there are two lines of approach.

Some people prefer a quick, definite change-over; others prefer a more gradual process, replacing products only as they run out. Do whatever you feel comfortable with. Your shopping will now cost a fraction of your normal bill. Less can mean more!

Bicarb soda (sodium bicarbonate, baking soda)

Bicarb soda occurs abundantly in nature but depends upon sophisticated processing in order to meet high standards of quality and purity. This non-poisonous powder is inexpensive and has a multitude of uses within the home. It will remove stains, absorb unpleasant smells, soften water, polish, relieve itching and act as a toothpaste and a deodorant.

Borax

Borax is a naturally occurring mineral salt. This fine white powder acts as a stain remover, grease solvent, natural deodorant, fabric and water softener, and soap booster. It possesses bleaching and disinfectant qualities as well. Borax is effective in controlling insect pests.

It must be recognised that borax is poisonous when swallowed, so care needs to be taken in its use and storage. It can also enter the body through broken skin.

Cloves

Cloves are the dried flower buds of tropical myrtles, used as spice. They are non-poisonous, sweet-smelling, readily available and inexpensive. Clothes moths find them totally unacceptable.

Eucalyptus oil

This natural oil is distilled from gum trees (eucalypts). It has many medicinal properties ranging from relieving the symptoms of colds and influenza to easing aches and pains. Eucalyptus oil is a penetrating oil that evaporates rapidly and is useful as an antiseptic, disinfectant and deodorant. It can be purchased as a spray or liquid.

As a cleaning agent, its uses range from freshening a load of washing; to removing grease, gum and stubborn stains from clothes; to lifting tar and adhesive material from paintwork and dogs' paws! It is also an important part of most wool washes and is useful as an insect repellent.

Although natural in origin, eucalyptus oil contains ingredients that are highly toxic when swallowed. Keep out of reach of children.

Fly-swats

One fly-swat for the kitchen, one for the laundry and you can get rid of those pest strips, aerosol sprays and insect bombs.

Laundry and kitchen detergents

There are two things to look for when purchasing detergents. Purchase brands that are:

- *Biodegradable* (capable of quick decomposition into ecologically harmless products).
- *Low phosphate* or *phosphate-free*. The problem of declining water quality in our inland rivers, lakes and waterways – due to algal blooms (particularly blue-green algae) – is a serious ecological issue. The most important cause of this problem has been identified as an excessively large quantity of plant nutrients entering

the waterways, especially phosphates. It has been estimated that 30 to 50 per cent of all phosphates that pass through sewage farms come from household detergents and other household cleaning products. You can play your part in helping to reduce this pollution by making small changes in your domestic routines.

Lemons

There's nothing more useful and attractive than a lemon tree in your home garden.

Due to its acid nature, lemon juice has many uses apart from as a food. Lemon juice serves as a mild bleach, deodorant, and cleaning and polishing agent. It will also soften stains and repel insects.

Olive oil

Pressed from the fruit of the olive tree, this oil is commonly used in cooking but has many other uses as well. As a protective oil with smoothing qualities, it is an excellent beauty aid. It is also good as a polish for wood and leather surfaces.

Pure soap

Dependable, durable and made of ingredients that are environment-friendly, pure soap is unlikely to cause skin irritations or allergies. It causes grease, food residues, bacteria and all other forms of dirt to disperse in water. The same brand of pure soap can be used in the bathroom, laundry and kitchen. It's also useful as a garden spray.

Salt

Salt is naturally occurring and non-toxic, and a very important seasoning and preserving agent for food. To use it as an antiseptic to clean cuts and grazes, simply dissolve half a teaspoon of cooking salt in 1 cup of boiled water. This solution is recommended by dentists as a mouthwash and to treat gum disease – and also after tooth extraction. For an eye bath, eye specialists recommend a weak salt-water solution.

Salt is also a valuable product when cleaning and disinfecting food preparation areas and utensils. To clean and disinfect a drain, simply use a handful of salt followed by a jug full of boiling water. Salt helps prevent colours running in the wash and removes stains. Its uses are widespread.

Soap shaker

Powered by your wrist, a metal soap shaker makes use of all those leftover slivers of soap and offers a cheap, safe alternative to detergents for washing your dishes.

Scourer (nylon) and steel wool (plain, fine)

Scourers are a simple, inexpensive and highly effective means of removing stubborn dirt without using powerful and potentially harmful cleaning products.

Vinegar (white)

White vinegar is an inexpensive, clear, colourless liquid made by fermenting cereals or cane sugar. This mild natural acid (acetic acid, with an acidity of about 4–5 per cent) is able to neutralise grease and soap residues. It is also an anti-mould agent, a mild disinfectant, a bleach and a deodorant. Vinegar is the very best general purpose cleaner and is environment-safe!

Washing soda (sodium carbonate, 'Lectric' soda)

This crystalline powder (or crystals) is useful as a water softener, stain remover, degreaser, silver tarnish remover and drain-cleaning agent. It can also be used as a poultice to reduce swelling, and in a bath to relieve aches and pains. Do not use on silk, wool, vinyl or aluminium.

Does this green shopping list seem too simple and inexpensive to be true? Well, I promise you that it's possible to keep your family and home sparkling clean using these basic items.

Green living begins in the home; your home.

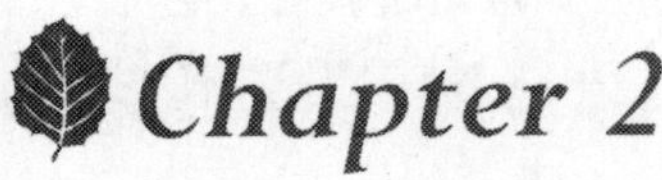

Chapter 2

It's Easy Being Clean and Green

Follow the green way of living and your home will be clean, pests will be reduced to a minimum and no cleaning task will ever seem too difficult to tackle. It must be remembered, however, that this is a guide only. No one lives in a home that operates with 100 per cent efficiency – and nor would we wish to.

DAILY TASKS

- Use a dry towel to wipe over wet areas (kitchen, laundry and bathroom). This will reduce the build-up of soap scum, bacteria and mould.
- Open up windows and doors to allow through ventilation. In addition, use an exhaust fan in your kitchen and bathroom. This will reduce mould and replace any harmful chemicals with fresh air.
- Sweep or vacuum the floor of your kitchen and eating area.
- Do an immediate wipe-up of food spills (and other dirty marks) on benches, tables and floors.

This way spills never spread or become difficult to remove.

- Check that food is correctly stored in your freezer, refrigerator or in sealed containers in your pantry. Stale food, food crumbs and residues will attract and feed bacteria, cockroaches, flies, ants, rats and mice.
- Develop a 'tidy as you go' habit. This way you'll have the pleasure of a well-organised home and you'll no longer spend large slabs of time trying to locate, for example, your car keys.
- Hang up your bath and hand towels to dry. Dryness kills both bacteria and mould.
- Check the toilets. Do they need to be brushed? Do the seats need a quick wipe-over with vinegar?
- Hang the kitchen sponge or cloth, scourer and dish mop outside in the sun. Sunshine, frost and dryness all kill bacteria and mould. Nature is the very best steriliser.
- Check that rubbish is correctly stored in sealed containers.

If cockroaches are a problem in your area, the following routine is especially important:

- During the hours of darkness, leave your kitchen and eating areas as clean as possible. Your aim is to provide as little food and water as you can for these unpleasant insects to feast on.
- In the morning, give your benches, sink and stove-top a quick wipe-over with white vinegar to ensure clean food preparation surfaces.

WEEKLY TASKS

- Wash bed linen and towels – and your pet's blanket.
- Dust throughout the house.

- Vacuum floors.
- Give a vinegar wipe-over to glass-topped surfaces, as well as to kitchen, laundry and bathroom surfaces – not, of course, extending to ceilings, walls and windows!
- Remove any stains on kitchen benches, stove, laundry and bathroom surfaces by sprinkling some bicarb soda over the stain, splashing on some vinegar, rubbing with a cloth, then rinsing clean. To give a final shine, wipe over the surface with a cloth moistened with vinegar. (With any luck this stain removal routine won't be required.)

These routines will keep house dust mites at acceptable levels, as well as mould and bacteria, ants, cockroaches, flies, silverfish, carpet beetles, rats and mice. Shared between members of your household, this routine is quite manageable.

USING GREEN CLEANERS

<table>
<tr><td colspan="2">Water
(hot, warm & cold)</td><td>Vinegar
(white)</td><td>Lemon
juice</td><td>Eucalyptus
oil</td></tr>
<tr><td>Soap
(pure)</td><td>Bicarb
soda</td><td>Salt</td><td>Borax</td><td>Washing soda
('Lectric' soda)</td></tr>
</table>

Working with a soft cloth, scourer or plain, fine steel wool, you can combine any of the liquids in the top row with any of the solids below.

These green cleaners increase in strength from left to right. Warm soapy water, for instance, is a very effective way of cleaning most surfaces. Stubborn stains sometimes require borax, 'Lectric' soda or eucalyptus oil.

- *Vinegar* can be used full-strength or diluted. For a general purpose cleaner, simply add 1 cup of vinegar to 4 litres of hot water.
- *Lemon juice* is used in the same way as vinegar.
- *Bicarb soda* is useful as a powder, paste (mixed with water or vinegar) or dissolved in hot water. To remove mould, dissolve 3 tablespoons of bicarb soda in 2 litres of hot water.
- *Borax* is effective as a powder, paste (mixed with water or vinegar) or dissolved in hot water. For cleaning dirty surfaces, dissolve 4 tablespoons of borax in 4 litres of hot water. For a general purpose solution to help remove stains on clothing, fabrics or carpet, use 1 tablespoon of borax for every 4 litres of hot water.
- *Washing soda* ('Lectric' soda) is used as a liquid. Dissolve 1 tablespoon of 'Lectric' soda in 1 litre of hot water.
- *Salt* crystals are an effective cleaning agent. Salt can also be useful as a solution. Dissolve 1 tablespoon of salt in 1 litre of water.

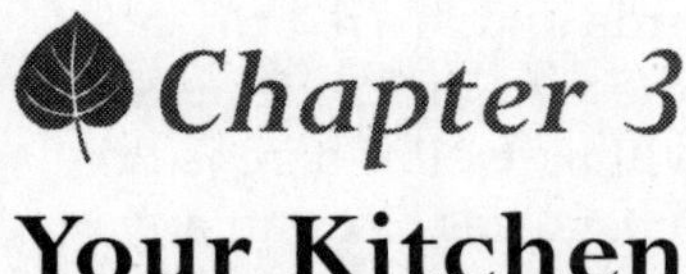

Chapter 3
Your Kitchen

By using green cleaning products, you are helping to protect our fragile environment. Even small changes in domestic routines will benefit the environment – and help spread the green message.

Harsh, expensive cleaning agents and insect sprays are unnecessary and should have no place in your kitchen. So, as you toss them out, concentrate your thoughts on improved air and water quality and smile because the air is beginning to smell so good! An efficient exhaust fan over your stove is an essential way to upgrade air quality. Cross-ventilation is also important.

Remember: sensible standards of cleanliness and ventilation are essential for good health. Using chemicals as a substitute for basic cleanliness and good ventilation is fraught with danger: danger for your health, danger for the environment. Hot water, soap, vinegar, bicarb soda and a little elbow grease – that's all it takes to be clean and green in the kitchen.

CONTROLLING BACTERIA

Television advertisements tend to cause unnecessary anxiety about bacteria, with talk of salmonella and E. coli. Viewers feel guilt about their housekeeping standards and are led to believe that if they spray these miracle cleaners around their kitchen, all will be well. There is no reason to use poisonous substances to clean any appliance or surface used in the preparation of food. Chlorine and ammonia-based bleaches, disinfectants and detergents are unnecessary in the home kitchen, often giving a false sense of security as regards bacteria. Their toxic vapours irritate delicate eye, nose and lung tissue and are often responsible for poor quality indoor air. Sensible hygiene needs to be practised rather than excessive use of anti-bacterial solutions that lead to the emergence of even more harmful strains of bacteria. Over-use can also lead to more childhood eczema, hayfever and asthma due to an under-stimulated, weakened immune system.

Your home can be kept modern, attractive and sparkling clean, with harmful bacteria kept at safe levels, by using environment-friendly products – plenty of soap and hot water, vinegar, salt and eucalyptus oil; as well as nature's sterilisers, frost and sunshine.

Washing your hands in warm soapy water is an extraordinarily important habit to adopt. Research suggests that ordinary soap eliminates harmful bacteria as effectively as anti-bacterial soap. The aim is not to sterilise, rather to reduce the number of harmful micro-organisms to a level where they will not cause illness. Ensure, therefore, that soap and a hand towel are always available in your bathroom and kitchen, and teach young children these golden rules. Always wash hands:

- After using the toilet.
- After handling pets.
- Before preparing food.
- And before eating.

CLEANING

Since the kitchen and the preparation of food are closely linked, a clean kitchen is essential. But this need not be at the expense of the environment.

Hot soapy water is sufficient, in most cases, to clean kitchen surfaces.

Bicarb soda is a gentle cleaning agent. It can be used as a powder, paste or solution. Stains on all types of surfaces can be removed using bicarb soda as a paste and applying it with a damp cloth. Rinse well, using hot water. Stainless steel surfaces, enamel, stained cups, refrigerators, china teapots, laminex surfaces and plastics all clean brightly with an application of bicarb soda.

If the stain is more stubborn, cover with a layer of bicarb soda, then sprinkle with vinegar. While bubbling, rub, using a soft brush, scourer or cloth. If you leave the bicarb soda and vinegar to sit for 10 minutes, the stain will soften. Alternatively, use bicarb soda moistened with lemon juice for extra cleaning power.

Vinegar is a mild acid that is able to neutralise grease and soap residues. It's also an anti-mould agent, a mild disinfectant, bleach, deodorant and general purpose cleaner. And it's environmentally safe, as well as being inexpensive, easy to use and effective. You can use it anywhere in your home: on kitchen benches and tabletops, windows, floors, bathroom surfaces, the toilet and in your rinse water for clothes and dishes.

Although the smell of vinegar disappears as soon as it dries (and it dries very quickly when there's plenty of ventilation), some people prefer to make their own herbal vinegar. You can use lemon balm, marjoram, any type of mint, thyme, lavender, sage, rosemary or any other herb you enjoy. Mint leaves give a fresh fragrance for cleaning, with the vinegar absorbing the aromatic qualities very well. To make your own herbal vinegar for cleaning, follow these easy steps:

- Pick 2 cups of leaves and put them into several pre-warmed bottles.
- Slowly heat 5 cups of plain label white vinegar, then pour the hot vinegar over the leaves.
- Allow to cool, then seal.
- Leave the bottles on your bench (preferably in warm sunshine), as you will need to give the bottles a shake every 2 days.
- Allow 2 weeks (you can leave it longer) for the vinegar to absorb the aromatic quality of the herb, then strain and reseal, ready for use.

Herbal vinegar will do everything that ordinary vinegar will do.

Salt is ideal for cleaning sinks and chopping boards used for food preparation, due to its excellent cleaning and disinfecting qualities. Glass, marble, metals and laminex also clean well using salt. Combine bicarb soda and salt to make a tough cleaning powder.

If you've become accustomed to using a commercial spray-and-wipe, you may like to pour undiluted vinegar into a spray bottle for use in a similar way.

Now, with the cupboards under your sink clean and uncluttered, and your stock of environmentally sound products on hand, here are some practical suggestions for using them.

Adhesive labels (to remove from bottles and containers)

Moisten label with eucalyptus oil and rub clean. Now wash with hot soapy water.

Alternatively, place your jar in the freezer and when the glue has frozen scrape it off using a knife.

Aluminium saucepans

Boil a solution of 1 part vinegar to 1 part water in the stained saucepan for several minutes.

Or, place any chopped-up acid fruit such as lemons, oranges, apples or rhubarb in the saucepan, fill with water and boil for 15 minutes. Give a final clean with steel wool, then rinse well and polish dry.

Baby's feeding equipment

Purchase bottles, dummies, teats etc. which can be boiled, then sterilise by immersing in boiling water.

Boiled-over milk

Turn off element or gas. Sprinkle generously with cooking salt. Leave for a few minutes. Wipe off with a damp cloth. The stove will now be sparkling clean and odourless.

Bone knife handles

To whiten and remove stains, rub the handles

with half a lemon dipped in salt. Rinse in warm water and dry with a soft cloth.

Bottles, glass decanters and vases

For bottles and vases with narrow necks, clean with steel wool attached to a thin stick. Rinse with hot water to which a little white vinegar has been added. Air dry.

Brass, bronze, pewter and stainless steel

Mix equal parts salt and flour. Now add enough vinegar to make a stiff paste and apply to the surface. Allow the paste to dry, rinse off quickly and buff dry with a soft cloth.

Or, apply bicarb soda on a damp cloth. Let it dry. Polish with a soft dry cloth.

Or, rub with a soft cloth moistened with vinegar. Polish with a dry cloth.

For brass curtain rings, simply drop the rings into a hot vinegar and water solution to soak. Buff dry.

Cake tins

Bicarb soda moistened with vinegar will remove unsightly stains from cake tins. Rinse with hot water.

Chopping boards

It is very important that chopping boards are thoroughly cleaned after use, especially after cutting up raw meat. Scrub plastic with hot soapy water. Rinse under hot running water, wipe over with vinegar, then allow to dry.

To clean wooden boards, rinse with cold water, then rub in plenty of salt. (Hot water opens up the grain and allows germs and odours to penetrate the wood.) When dry, season the wood with a little vegetable oil. Keep one chopping board for raw meat only, another for vegetables to be cooked, and yet another for salad fruits and vegetables.

White polythene chopping boards can be rubbed clean using a paste made up of equal parts salt and bicarb soda, with water. For bad stains, add lemon juice to the paste rather than water.

Since it's non-porous, hardened glass makes a good food preparation surface. It's easily cleaned with vinegar – just like a window.

Copper pans and copper-based pans

For a brilliant sheen, dip half a lemon in salt and rub the surface vigorously. Rinse with hot water. Polish dry with a soft cloth.

Or, moisten salt with vinegar and rub into the copper surface. Rinse with hot water. Polish dry with a soft cloth.

Dishwasher

To clean the dishwasher (of scum and mineral deposits), add 1 cup of vinegar to the dishwasher and set the machine for a normal cycle.

Dish washing

Water is the cheapest and most effective cleaner available and, when hot, is usually sufficient to wash dishes. Use pure soap in a shaker if your water is soft and your dishes are a little greasy. If your water is hard, an environmentally safe detergent will be needed for greasy dishes.

A little vinegar added to the rinsing water gives dishes, cutlery and glasses extra sparkle by removing all traces of soap and detergent. Give the sink a final wipe-over with vinegar to give a clean, streak- and smudge-free surface.

Assuming your kitchen is free of flies and dust, the most hygienic way to dry dishes is to leave them on a draining rack.

A final hint: develop the daily routine of hanging your dishcloth, scourer and mop in the sunshine to dry, and let nature do the sterilising.

Drains (to clear blockages and disinfect)

Commercial drain cleaners contain caustic soda, which is highly corrosive and toxic if swallowed. Fumes may cause serious eye, skin and throat irritation. Here are some safe, effective alternatives – safe for the environment, safe for your health.

A handful of salt followed by a jug of boiling water does an excellent disinfecting job. Half a cup of vinegar helps to remove any unpleasant smell.

For extra-stubborn blockages, pour 1 cup of washing soda and half a cup of vinegar down the drain. Let it sit for half an hour with the plug in place. Then give a final salt and boiling water treatment. Washing soda ('Lectric' soda) is especially good at dissolving grease.

As a preventative measure, use a sieve that fits into the drain hole in your sink to collect any 'stray' material.

Egg stains

Wet salt or a bicarb soda paste are both excellent for removing egg stains from crockery and cutlery.

Electric frypan

To remove stains and grease from the outside, clean with a pad of steel wool soaked in eucalyptus oil or methylated spirits. Rinse clean with very hot water.

Freezer (inside)

Use a solution of 3 tablespoons of bicarb soda to 6 cups of warm water to clean and freshen.

Fruit and vegetable stains (on hands)

Put 1 teaspoon of sugar in the palm of your hand. Rub the inside of a lemon skin into the sugar then over the stains. Rinse hands clean.

Glasses (drinking)

Those used for special occasions often need a little attention prior to use. Partly fill your kitchen sink with hot water, then add half a cup of white vinegar. Let your glasses soak for a few minutes, then take out and allow to drain on a clean tea towel. While they are still hot, polish dry with another clean tea towel. Your glasses will now sparkle, free of dust, streaks and smudges.

Granite bench tops

To clean, take a freshly cut lemon, dip its cut edge into warm water and then salt, and rub the surface with this lemon pad. Alternatively, use bicarb soda moistened with vinegar, on a cloth. Rinse clean.

To polish, sprinkle with vinegar, wipe over the surface, then buff with a soft dry cloth.

To remove stains, mix equal parts bicarb soda and salt with water to the consistency of thick cream.

Smear this over the stain and leave overnight. Rinse clean.

Kettle (to remove lime deposits)

Lime deposits can be removed by boiling a solution of equal parts vinegar and water. If the mineral deposit is particularly bad, allow the kettle to stand overnight before rinsing clean. Alternatively, cut a lemon into small pieces, then place the lemon pieces inside the kettle. Fill with water and boil. Leave overnight, then rinse well the following morning.

Knife blades

Knives can be laden with invisible bacteria, so be sure to rinse your knife under hot running water whenever you change from one type of food to another. Using the same knife to cut raw meat, then slice salad greens and tomatoes, is very risky in terms of the transfer of harmful bacteria.

Laminex

Sprinkle vinegar over the laminex, then wipe with a damp cloth to produce a fresh-smelling, shiny, clean surface.

Stains may be removed, without scratching, by combining bicarb soda with water (or vinegar) and rubbing clean. Sponge off the bicarb soda, then finish the job with a vinegar wipe-over.

Microwave oven

After use, wipe the inside with a hot damp cloth, especially if there has been a spill.

For more stubborn stains, place a bowl of hot water (containing a slice of lemon) in the microwave. Set to simmer until the interior is quite steamy. Leave for a few minutes, then remove the bowl. Now wipe away the stains using a hot damp cloth.

Oven

Research suggests that exposure to commercial oven cleaners adversely affects the production of healthy sperm. However, this doesn't mean that cleaning the oven is women's work! Most oven cleaners produce toxic vapours which can (and often do) cause headaches, rashes, sneezing, watery bloodshot eyes and fatigue. There's no need for anyone to risk their or their family's health using these substances. There are safe, effective alternatives.

While your oven is still warm, wipe with a damp cloth sprinkled with bicarb soda.

If the oven is really grimy, apply a bicarb soda paste to all surfaces, then leave for 30 minutes. Now wipe clean. Use a wooden scraper on any difficult patches. Add vinegar or lemon juice to the bicarb soda for extra cleaning power.

Soak oven racks in hot soapy water and bicarb soda. Polish clean with steel wool, rinse with very hot water and dry with a soft cloth.

To prevent a grimy oven, make up a mixture of 1 tablespoon of bicarb soda dissolved in 1 cup of water. After cleaning, 'paint' your oven with this mixture. When dry, the bicarb soda and water will leave a hard sterile skin over the clean surface. Any fat, oil or burnt food will now stick to the rough surface made by the bicarb soda. The bicarb soda and grime will lift off easily using plain hot water. Every month or so, recoat all the surfaces again. This way you can be certain you won't serve up a film of harsh oven cleaner along with your next baked meal!

When you clean with bicarb soda and vinegar, you're using cleaning products that are so safe you can and do eat them. Besides being safe and inexpensive, these products will not go on to pollute our waterways and oceans.

Oven door (glass)

Clean with dry bicarb soda sprinkled on a damp cloth.

Oven (food spill)

Sprinkle with salt, then brush off the burnt food which will have combined with the salt.

Painted surfaces

Wash with hot soapy water, or sponge with vinegar on a moist cloth, or bicarb soda on a damp cloth.

Pastry board

Use a knife to scrape off dough, then sprinkle board with salt and rub clean with a cool damp cloth.

Plastic electric jug and kettle

Moisten bicarb soda with vinegar and apply with a soft cloth. Rinse well and polish dry.

Since plastic is soft as compared with metal, it's wise to select metal percolators, electric jugs, kettles, toasters and so on. This will reduce your intake of unwanted plastic particles.

Plastic ware

Use bicarb soda dampened with water to clean

stains off plastic cups, plates and bowls. This will not damage the surface.

Refrigerator (inside)

To keep your refrigerator sweet-smelling and free of mildew, wash the surfaces with a solution of bicarb soda and hot water, or sponge with vinegar on a damp cloth.

Roasting dish (burnt fatty food)

Spread a sloppy paste of bicarb soda over the burnt food and leave the dish to soak overnight. Next morning, wipe off the paste, which will lift the burnt food with it. Rinse clean with very hot water.

Rubbish bin

Use plenty of hot water, with a little detergent and eucalyptus oil added, to clean and disinfect.

Saucepan (burnt food stuck to bottom)

Cover with a sloppy paste of bicarb soda. Leave overnight and rub clean the following morning.

Or, apply bicarb soda and lemon juice paste, and leave to soak.

Or, fill saucepan with cold water, then add half a cup of salt. Soak overnight. Bring slowly to the boil and let it simmer for 5 minutes. Leave to cool. Drain off the liquid and polish clean.

Or, cover with a salt and vinegar paste and leave overnight. Add a little water in the morning, then bring slowly to the boil. Simmer for 5 minutes. When cool, rub clean with steel wool.

Saucepan (to prevent staining)

When boiling spaghetti, rice or eggs, add half a

teaspoon of vinegar (or a little lemon juice) to the water to prevent staining.

Silver (tarnished)

Place silver in an old aluminium saucepan. Cover with water. Add 3 teaspoons of washing soda (or 3 teaspoons of salt or 3 teaspoons of bicarb soda). Bring to the boil. Simmer for 3 minutes. Remove silver, rinse in hot water, then let drain on a soft cloth. Polish each piece dry with a second cloth. An old toothbrush can be used to help remove tarnish from crevices.

If you haven't got an aluminium saucepan, simply put a piece of aluminium foil in any other saucepan or line the bottom of the kitchen sink with aluminium foil and fill the sink with very hot water, then follow the hints above. The tarnish is attracted to the foil.

Silver cutlery can be kept shiny by storing it in a specially made cloth bag that has individual pockets for each piece.

Soap residue (in bottles)

Rinse with vinegar and hot water.

Stainless steel cutlery (to make it sparkle)

Add 2 tablespoons of bicarb soda to a jug of boiling water and soak cutlery for 10 minutes. Rinse well and polish dry.

Tea and coffee stains

Rub a bicarb soda and water paste into stains on china, ceramic or plastic cups and mugs. Rinse off and dry with a soft cloth. Bicarb soda will not scratch the surface.

Teapot

Sprinkle salt into the teapot and rub with a damp cloth. Rinse with very hot water.

Or, fill the teapot with warm water, then add 1 tablespoon of bicarb soda. Let it stand for 10 minutes, then clean with a bottle-brush.

Thermos

Place 1 teaspoon of bicarb soda and warm water in the flask and allow it to soak for a few minutes. Rinse well.

Vegetables

To help remove pesticides, bacteria and traces of mould or mildew from vegetables (both salad vegetables and vegetables to be cooked), combine a little vinegar with water and soak fruit and vegetables for a few minutes. Brush clean.

Vinyl (scuff marks on)

Use eucalyptus oil to rub away the marks.

Vinyl floors (also ceramic, cork, slate, terracotta and vinyl tiles)

Mop using a solution of 1 cup of vinegar to half a bucket of very hot water, or rub away stains and other dirty marks using a cloth moistened with vinegar. The vinegar cleans, deodorises, is an anti-mould agent, prevents spotting as the floor dries and leaves a shiny surface. Take care not to overwet the surface, especially when cleaning cork or vinyl tiles.

Maintaining a sterile floor surface is a waste of time and resources. Food spills and dirty marks do need to be cleaned up, however, and eating area and kitchen floors swept or vacuumed daily.

DEODORISING

Chopping board

Half a lemon rubbed into the surface will clean the board of onion, garlic and fish odours. Alternatively, rub with a handful of fresh parsley or mint.

Dishcloth

Soak a smelly dishcloth (the smell indicates the presence of bacteria) overnight in a solution of vinegar and hot water. Rinse, then hang in the sunshine, a drying wind or frost and let nature do the deodorising and sterilising.

Fish (cooking smells)

A bowl of vinegar placed beside the pan will help reduce the smell of hot fat and oil.

If the smell is still strong, add 1 tablespoon of vinegar to the washing-up water.

Hands

To remove unpleasant or strong odours from your hands as a result of preparing onions, fish etc., use any of the following methods.

Rub hands with salt and water, or bicarb soda and water.

Or, work half a lemon into your skin.

Or, rub your hands with vinegar (or herbal vinegar), rinse, then wash in warm soapy water.

Or, massage a couple of drops of eucalyptus oil

into your skin, then wash with soap and warm water.

Or, crush a handful of parsley between your fingers and palms.

Microwave oven

Add the juice of 1 lemon to half a cup of water. Place in microwave and turn on high setting for 1 minute. Wipe out. Repeat if necessary.

Remove strong odours produced by foods such as pickles or chutney by wiping over the oven's interior with equal parts vinegar and hot water. Leave the door open overnight as well.

Oven

When you clean your oven using bicarb soda and vinegar, any lingering smells will be neutralised, especially if you leave on the recommended light coating of bicarb soda and water.

With the oven off, you can place a small dish of vanilla essence on the bottom shelf to deal with really unpleasant odours.

Plastic container

To remove unpleasant odours, soak in bicarb soda and warm water. Rinse well.

Alternatively, use lemon juice and water, or vinegar and water, or herbal vinegar and water. Wipe dry, then store with the lid slightly ajar.

Plastic drink bottle

Soak overnight in a solution of bicarb soda and warm water to remove both the taste and smell of the plastic.

Plastic electric kettle

Pour 1 cup of white vinegar into the kettle, then fill with tap water. Allow the mixture to boil. Rinse well.

Refrigerator

To mask unpleasant odours in your refrigerator:

Place a small dish of vanilla essence on the bottom shelf.

Or, place half a lemon on a shelf to help absorb food odours.

Or, leave an open packet of bicarb soda in the refrigerator. This will be effective for about 3 months.

Rubbish bin

A light spray of eucalyptus oil helps to neutralise disagreeable smells.

Thermos

Store a piece of charcoal in your thermos to keep it smelling fresh. Remove before use.

Unpleasant odours of all kinds

To freshen up your kitchen, open up those windows! Fresh air is both free and the very best deodorant. However, if nature needs a little help, use any of the following in a small bowl: vinegar; herbal vinegar; cloves; activated charcoal; fresh herbs or sweet-smelling flowers; any of the fragrant oils, such as eucalyptus, tea tree or citronella; or boil a handful of cloves in a saucepan of water for a few minutes.

Alternatively, place an open packet or bowl of dry bicarb soda in the area to absorb smells for about 3 months.

Vinyl and plastic surfaces (new)

Wash with a strong solution of vinegar and hot water, then dry using plenty of ventilation. New kitchen floors (also car interiors) benefit greatly from this treatment.

As you can see, every odour problem can be solved using something safe from the average pantry.

MOULD REMOVAL

Mould produces microscopic spores and potentially harmful vapours that float in the air. When they are inhaled or swallowed or land on the skin, they cause allergic reactions in some people. The allergic reaction can lead to asthma, eczema, sinusitis, conjunctivitis or rhinitis, with hayfever-type symptoms. Apart from the health problems triggered by mould and mildew, there is also the unsightly aspect to contend with. The following green hints will help you with this problem.

White vinegar, lemon juice, salt, Epsom salts, eucalyptus oil and tea tree oil can be used to kill and clean away mould and mildew. Simply wipe over the surface using a strong solution of any of these products, leave for 10 minutes (or overnight) then wipe clean.

If the mould is particularly ingrained, combine bicarb soda with vinegar and use a stiff brush to clean the area – and if possible, use sunshine to dry.

When using eucalyptus or tea tree oil (both oils have anti-fungal properties) to clean away mould and mildew, add a few drops to hot soapy water, then use a vinegar and hot water rinse.

Eucalyptus spray is useful as a preventative measure.

Remember that mould and mildew thrive in warm, damp, poorly ventilated places, so use an extractor fan and allow good through ventilation.

Bread containers

Mix 1 tablespoon of vinegar to 2 cups of hot water and wash thoroughly. Dry the bread container in the sun.

Painted surfaces

To clean mould from walls, ceilings and so on without the use of expensive bleaches and harsh cleaning agents, you will need:

- Four containers, clean cotton cloths, bicarb soda, vinegar and hot water.
- Container *one* holds a mixture of bicarb soda and hot water (3 tablespoons bicarb to 2 litres water).
- Container *two* holds a mixture of vinegar and hot water (half a cup of vinegar to 2 litres water).
- Containers *three* and *four* hold hot water only and are used to rinse the cloths between use. Remember to rinse the bicarb soda and vinegar cloths in different rinsing buckets.

First use the bicarb soda solution, then the vinegar.

If a patch is particularly difficult to remove, bicarb soda can be used dry, as an abrasive. Alternatively, moisten salt with lemon juice and rub into the stain. Leave for 30 minutes then rinse well and dry.

Refrigerator

Wipe all surfaces with a hot vinegar and water solution. For the sealing strips around the door in particular, add a few drops of eucalyptus oil to 1 cup of very hot water, then wipe over the sealing strips. Using this method, you can control mould and mildew.

OTHER GREEN KITCHEN HINTS

Brown paper bags

Fruit ripens very well stored in brown paper bags at room temperature.

Vegetables last well stored in brown paper in the refrigerator.

Cellophane bags

Cellophane, made from wood fibre, is a safer product to have in contact with food than plastic. It is a natural, biodegradable product, and many organically grown products are packaged in it. Cellophane bags can be wiped clean and re-used many times. They are ideal for use in the freezer.

Greaseproof paper and brown paper bags are safer for packed lunches than plastic wrappings.

Soap scraps

Place soap scraps into a soap shaker and use to wash dishes.

Steel wool

To prevent steel wool rusting, place the used pads in a screw-top jar in which 1 teaspoon of bicarb soda has been dissolved in warm water. Alternatively, use a soapy mixture.

Robin's favourite green hints

On my way to the clothes line, I pass a collection of herbs that tumble over brick paving. I love the smell of crushed rosemary and thyme, and often pause to pick a sprig of parsley to rub between my fingers.

Laminex benches receive the vinegar treatment, which means I splash vinegar about, then wipe over all surfaces using a hot squeezed-out cloth. If they're stained, I sprinkle bicarb soda over the stain, splash vinegar on to the bicarb soda and, while fizzing, rub with a cloth or brush. Then I wipe the laminex clean, rinse and give it a final vinegar wipe. There are no fumes to irritate my eyes, nose and lungs, and there's no need to wear rubber gloves.

In each of the kitchen, laundry and bathroom cupboards, I have a container of bicarb soda and a bottle of white vinegar. Each bottle of vinegar has three holes drilled in its plastic lid. This means that I can easily dribble vinegar over any surface, then wipe clean for a shiny, streak-free surface that is disinfected and deodorised as well.

Our copper-based saucepans receive the cut lemon and salt treatment, and I'm always amazed at how brilliantly this works!

Rinsing the dishcloth in hot water and vinegar is my way of freshening it up. The dishcloth, scourer and dish mop hang out in the wind and sunshine as often as possible, which is most days.

Tea and coffee stains come away easily with bicarb soda on a damp cloth.

All our salad vegetables, as well as those that are cooked without peeling, are rinsed in a weak vinegar and water solution. This ensures that almost all pesticide residue, bacteria, mould and mildew are removed.

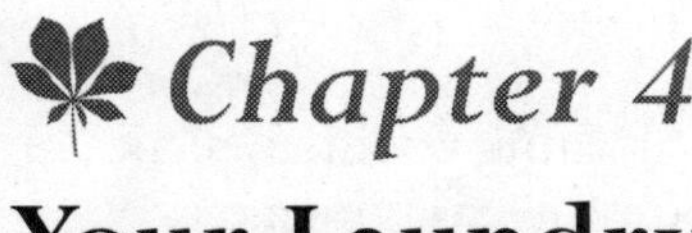

Chapter 4
Your Laundry

The contamination of our environment has become an important issue, and one which concerns every thinking person. You may think that the quantity of contaminants discharged from your household is small, but the cumulative effect of millions of households is drastic. By making small changes in your domestic routines, you can significantly reduce the pollution in our rivers, lakes and oceans.

As responsible members of the community, it's up to us to ensure that our waste water is as biodegradable as possible and contains no phosphates. We can feel confident of this if we use the following hints and products.

Bleach

To whiten clothes, add half a cup of borax to a normal wash cycle, then hang clothes in the sun to dry.

To bleach the natural way, leave clothes out overnight in frosty weather. This is an excellent way of restoring crisp whiteness to table napkins.

Lemon juice is a safe and effective bleach. Simply soak in lemon juice and water.

White vinegar is a mild bleach. Soak overnight in a vinegar and water solution.

Colours (running)

To 'fix' the colour in new clothing or linen, soak the article in a salt-water (or a weak vinegar) solution for 2 hours before a normal wash.

Salt in the washing water will help prevent colour 'running' out of the material.

To fix a dark dye, soak the garment for 30 minutes in a solution of 1 tablespoon of Epsom salts dissolved in 2 litres of cold water. Then wash in the usual manner.

Colours (to enhance)

Add a quarter-cup of vinegar to the final rinse for cottons.

Add a pinch of salt to the washing water for blue fabrics.

Add a quarter-cup of vinegar to the washing water for black, navy, brown, fawn, bone, red and pink materials.

Eucalyptus oil

Add a few drops to any load of washing that needs a little extra cleaning and freshening.

Fabric softeners

Vinegar acts as a fabric softener. Soak the garment overnight in a solution of 3 parts water to 1 part vinegar, then wash as normal.

Bicarb soda has fabric-softening qualities as well. For your normal wash, use less soap and make up the difference with bicarb soda (3 parts soap to 1 part bicarb soda), and clothes will feel softer.

Borax is an excellent fabric softener. Add 1 tablespoon of borax to your wash, then proceed with a normal wash cycle.

Epsom salts soften fabrics as well. Dissolve 1 tablespoon of Epsom salts in hot water and add to your normal wash.

Felt hats (Akubra)

To remove spots, rub lightly with fine sandpaper.

Or, sponge or spray with eucalyptus oil to remove oil, dirt or grease stains.

Foam

Too much foam, from using too much soap powder, can be settled with the addition of a little salt or a splash or two of vinegar.

Hard water

Hard water can be softened with the addition of half a cup of bicarb soda (or borax or washing soda) to the wash cycle. Add vinegar for the rinse cycle.

Iron care

Distilled water for your steam iron may be obtained by collecting rainwater. This water is best stored in a glass bottle.

To clean the inside of your steam iron, pour equal parts vinegar and water into the iron and let it steam for several minutes. Rinse with fresh water.

To clean the surface of your steel-based iron, unplug and wipe with bicarb soda on a damp cloth. Or, wipe with a solution of hot vinegar and salt while the iron is still warm. Alternatively, rub the stain with steel wool moistened with vinegar, with the iron warm but unplugged.

To clean the surface of your teflon-coated iron, warm the iron then unplug. Now wipe with a soft cloth moistened with a little olive oil. Then give a final wipe with methylated spirits to remove the oil.

Lace

Old lace may be cleaned by soaking in a warm borax solution, then rinsing carefully and drying on a soft towel. Freshly laundered lace needs to be stored in tissue paper.

Linen (to iron)

Place the linen garment in a plastic bag in your freezer until very cold. Ironing will now be much quicker and easier.

Musty sheets

The cupboard used to store your linen needs to be well-ventilated, rather than airtight, to avoid that unpleasant musty smell. Cloves or bay leaves give a pleasant aroma that discourages stale smells as well as moths.

To treat musty linen, add 1 cup of white vinegar to your rinse cycle, then dry in the sun.

New clothes and bed linen

Bed linen and many new clothes come with instructions to wash prior to use. Washing is necessary because of formaldehyde, a chemical used to make fabrics shrink-proof, grease-resistant, dye-fast, flame-resistant and moth-proof, as well as to give 'body' and sheen to the material.

Formaldehyde is a potent irritant that may cause watery, sore eyes, nose irritation, sore throats, skin rash, headache or asthma. While browsing in the material or clothing section of any large store, many of us notice the unpleasant smell of dyes and chemicals such as formaldehyde – and the way these make our eyes, nose and throat feel.

To overcome this problem, soak new fabrics in bicarb soda and warm water, preferably overnight; then wash with soap, adding a quarter-cup of vinegar to your final rinse water. Dry out of doors, letting the sun and wind work their own magic, then finish off the job with your iron.

New jeans

To soften new jeans, soak overnight in a borax solution. Alternatively, an Epsom salt solution may be used. Follow with a normal wash cycle, adding vinegar to the rinse water.

New towels

New towels need to be washed before use. Soak overnight in a bicarb soda (or salt) solution, then put through a normal wash cycle, adding vinegar to the rinse water.

Rinsing

To remove all traces of soap or detergent, add half a cup of vinegar to the rinse cycle.

Half a teaspoon of eucalyptus oil will ensure a fresh smell, help repel moths and silverfish, and assist in house dust mite control.

Silk

Hand-wash in lukewarm water using pure soap. Add 1 dessertspoon of vinegar to the rinse water to preserve the natural sheen. Finish off with a 'just' warm iron.

Smoke (tainted washing)

If a neighbour's wood fire has left your washing with an unpleasant smoky odour, put the articles back in your washing machine. Now add hot water with a generous amount of white vinegar. Let the articles soak for 15 minutes, then spin dry. All traces of smoke will have disappeared.

Sneakers (runners or white shoes)

Sprinkle bicarb soda on a moist cloth and rub away stains. Rinse with another damp cloth, then polish dry.

Soft toys

Non-washable soft toys can be cleaned by dusting bicarb soda (or cornflour) into the material, then brushing clean.

Stainless steel (troughs, taps)

Apply bicarb soda on a damp cloth. Rinse with hot water. If you require only a quick wipe-over, use vinegar to produce a sparkling surface free of streaks and spots.

Starching

Use 1 teaspoon of borax to 4 tablespoons of starch. Prepare the starch in the usual way, then dissolve the borax in a little boiling water and add to the starch. The addition of borax will produce a pleasing gloss and preserve the natural texture of the material.

Washing machine

Detergents can be removed from your washing

machine with the addition of half a cup of bicarb soda to a normal wash cycle.

Hoses can be cleaned by the addition of 3 cups of vinegar to a full machine load of warm water. Let the machine run a full cycle and repeat every few months to save costly repair bills.

White streaks

The addition of half a cup of vinegar to a machine load of dark cottons and socks will help prevent ugly white streaks forming.

Woollen garments and blankets

Borax preserves the natural softness of wool. To wash, add 1 tablespoon of borax and a little eucalyptus oil to 4 litres of warm soapy water and gently squeeze the suds through the woollens. Rinse well, spin, then dry, making use of a warm wind.

Rinsing is essential to remove soap (or detergent) film which causes fibre weakness, faded colours and yellowing. The rinse water needs to be lukewarm and should contain a little eucalyptus oil as well as half a cup of white vinegar to improve the feel and neutralise any soap scum. The eucalyptus oil will also help repel clothes moths and silverfish.

Woollen sweater (shrunk)

Dissolve 1 cup of Epsom salts in 1 bucket of boiling water. Allow the solution to cool. Soak the garment for half an hour. After soaking, squeeze out the solution and stretch the sweater into its correct shape. When it's almost dry, iron under a dry cloth.

Zippers

Close zippers before washing.

If a zipper jams, rub soap or candle wax over the 'teeth'.

Robin's favourite green hints

Taking freshly washed bed linen out into the sunshine to peg on the line is always a pleasure. So too is bringing it back indoors. I love smelling the sunshine in sheets and pillow slips, especially when I transfer them directly from the clothes line to our bed.

Since I'm not fond of ironing, I give the clothes a good shake before hanging, and peg and unpeg carefully. I fold directly into my cane basket. This way I minimise the amount of ironing.

Vinegar is often used in our rinsing water to remove all traces of soap.

Wood ash was traditionally used by the Greek community to soften water used for washing. A kilo or more of ash was placed in a cloth on top of the family wash. Boiling water was poured slowly through the ash, then the clothes were left to soak overnight.

To remove stains, the Greeks had a traditional hint that works just as well today. Simply pick a few sprigs of rosemary and boil them in a little water for about 5 minutes. Take out the rosemary, then wet a white handkerchief in the liquid. Squeeze out excess moisture. Place the handkerchief over the stain and iron. The stain will lift, leaving the material looking clean and smelling fresh.

Chapter 5
Laundry and Carpet Stains

There are many, many green ways to tackle stains. The following points will help solve this age-old problem:

- With a fresh stain, the rule is *don't let it dry out completely.* Some dry stains prove impossible to remove no matter what you do, especially those caused by meat juice, blood, fruit or egg white. Therefore, aim to treat stains while they are fresh. When time is short (or you don't want to make a fuss), simply sponge with cold water, soda water or a white vinegar and cold water solution, then leave a damp towel over the stain until you can deal with it properly.
- Due to the wide variation of materials used in the manufacturing process, it is always wise to test the solution being used on a small piece of the fabric.
- To avoid a ring developing, first treat the area around the stain, then work back towards the centre.
- Remove stains before putting articles through a normal wash cycle.
- With carpet, avoid over-wetting.

Bicarb soda softens fabrics (use 3 teaspoons for every litre of warm water) and helps remove stains (3 tablespoons for every 2 litres of hot water).

Borax acts as a soap booster, grease solvent and mild bleach and will dissolve most dirty spots. Wash clothing, carpet or other fabrics in a borax and warm soapy water solution (use 1 tablespoon of borax in 4 litres of hot water).

Eucalyptus oil or spray can be used to remove biro, chewing gum, grass, grease, gum, glue, ink, lipstick, nicotine, oil, shoe polish, tar or any stubborn unknown stain. Simply place an absorbent cloth under the stain, then dab or spray with eucalyptus oil, working towards the centre of the mark. Follow with a normal wash.

Citrus-based cleaners, which use a blend of natural extracts and no phosphates, can also be used successfully on a wide range of stains.

Methylated spirits can be used to remove biro, felt pen, grass, grease, nicotine or shoe polish.

Washing soda (sodium carbonate), sold as 'Lectric' soda, softens water and removes stains, and is excellent at removing grease. For best results, dissolve crystals in hot water, then leave greasy articles to soak before putting through a normal wash cycle.

We all know the feeling of panic when that first drink is spilled over a favourite tablecloth or brand-new carpet! In general terms, a largish spill of any liquid requires a quick application of absorbent paper or towelling to soak up the fluid as fast as possible. On carpet, it helps to walk on the towelling to maximise absorption.

If your carpet (or upholstery) has a slight overall greasiness or odour, you can dry-clean it yourself. Simply fill a sieve with bicarb soda (or cornflour), then sprinkle the powder through the sieve and onto your carpet or rugs. Using a soft broom, brush the powder into the pile. Leave it overnight, or for at least half an hour. Finally, vacuum thoroughly and you'll find that the bicarb soda has absorbed all the odour, grease and loose dirt. Steam cleaning (without the use of chemicals) is the best solution for some people.

Stains are unlikely to become 'stubborn' if they are rinsed out or sponged straightaway with cold water, soda water, or white vinegar and cold water. However, if they do go unnoticed and become difficult, here are some things to try:

- Sprinkle bicarb soda over the stain, then moisten it with a little soda water and rub gently. If applying to carpet, do not over-wet. Allow to dry, then vacuum clean.
- Spray the spot or stain using eucalyptus oil. Dab clean using an absorbent cloth. Alternatively, soften with eucalyptus oil, then cover with a bicarb soda and water paste. Leave for 1 hour. Wipe clean using a soft absorbent cloth.
- Dust with borax (not beyond the edge of the stain), then dampen with droplets of water. Let it dry, then brush clean. Alternatively, cover stain with a borax and water paste.
- Cover with a bicarb soda and vinegar paste, but not beyond the edge of the stain. Allow to dry. Brush clean.
- Dab with or soak using undiluted white vinegar.

- Rub stain using a combination of bicarb soda or borax, an environment-friendly detergent and a little eucalyptus oil. Add vinegar to your final rinse water.
- Use a cleaning product made from natural citrus extracts.
- Use 'Lectric' soda or lemon juice to soften the stain.
- Try methylated spirits if all else fails.

If the stain remains, ask yourself the question, 'Does it really matter?'

Beer

Dab with vinegar, then rinse well using cold water.

Beetroot

Stains on a cotton or linen cloth may be removed by soaking in cold salty water.

Or, soak a slice of bread in cold water. Place the bread over the beetroot stain until all the colour is absorbed by the bread. Follow with a normal wash.

Bird droppings

Scrape off excess when dry, then sponge with a vinegar solution.

Biro

Sponge with a cloth moistened with eucalyptus oil. Alternatively, spray with eucalyptus oil, then soak in lemon juice prior to a normal wash. You may need to repeat this process.

Or, sponge with a cloth dipped in methylated spirits.

Blackberry

Sponge with a borax or vinegar solution.

Blood

If fresh, sponge or soak immediately using cold salty water.

If old, make a borax and water paste. Pat this onto the stain and allow to dry. Brush clean.

Burn mark

Remove burnt fibre with sandpaper, steel wool or nail scissors. Then sponge with white vinegar.

Candle wax

Use ice to harden the wax, then scrape it off with a knife. Now place a piece of absorbent paper (plain brown or blotting paper) both under and over the stain. Use a warm iron to melt wax and transfer it into the paper. Repeat with fresh layers of paper until the stain disappears.

Alternatively, place tablecloth in a plastic bag in freezer, then scrape off hardened wax. Now dab with eucalyptus oil to eliminate last traces of wax.

To avoid wax drips, refrigerate candles prior to use.

Cement

Combine 1 tablespoon of salt with 1 cup of vinegar and add to cold water. Soak cement-stained clothes in the mixture, then follow with a normal wash.

Chewing gum

Place the article in a plastic bag, then put it in the freezer for 2–3 hours. Alternatively, place ice in a cloth, then rest against the gum. Scrape off the hardened gum with a knife. Sponge with eucalyptus oil to remove completely all traces of the gum.

Chocolate, cocoa and coffee

Soak immediately in cold water, then sponge with a borax or vinegar solution.

Collars

Spray dirty shirt collars with eucalyptus oil, or cover with bicarb soda moistened with vinegar or a soap gel. Leave to absorb, then follow with a normal wash.

Cream

Scrape off surface cream, then sponge using cold water. Soak in a borax solution.

Curry

Sponge with a borax solution.

Egg

Sponge using cold salty water. Rinse well.

Felt pen

Dab clean with methylated spirits or equal parts methylated spirits and vinegar. Alternatively, spray with eucalyptus oil. Soaking in lemon juice will also soften the stain. You may need to repeat the process.

Fruit juice

Sponge with a cold borax solution.

Or, sponge with a cold bicarb soda solution to neutralise the acid.

Grass

Soap and warm water will usually remove grass stains from cotton. If they don't, dampen the stained area with water, then sprinkle with white sugar. Roll up and leave for 1 hour. Follow with a normal wash.

Or, sponge with methylated spirits or eucalyptus oil. To prevent a ring forming, place an absorbent cloth under the stain and work towards the centre of the mark.

Grease (or oil)

To clean grease (or oil) stains from any porous surface such as carpet, fabric, leather, bricks, concrete or stone, follow this procedure:

- Absorb surface grease with paper towelling. Alternatively, sprinkle with cat litter (white clay particle type) or dry sand, then sweep clean.
- Combine equal parts chalk and bicarb soda with enough water to make a sloppy paste.
- Spread paste over the stain. Seal with plastic.
- Leave until completely dry.
- Brush (or vacuum) away the chalk and bicarb soda, which will have absorbed the oil or grease. Repeat if necessary.

Or, pour boiling water through the grease spot, then dust liberally with bicarb soda. Work the bicarb soda into the stain, then wash normally.

Or, clean using eucalyptus oil.

Greasy clothing

Add 1 teaspoon of eucalyptus oil plus 1 tablespoon of 'Lectric' soda to the normal washing powder used in your machine. Allow to soak, then proceed with a standard wash cycle.

Handkerchiefs

To dissolve mucous material, soak in cold salty water prior to normal wash.

Ice cream

Scrape off as much as possible, then rinse in borax and cold water.

Ink

Use blotting paper to soak up as much ink as possible. Cover the spot with salt, then moisten salt with equal parts vinegar and water. Leave for 2 hours. Brush off salt and sponge with cold water.

Or, soak in borax and warm soapy water.

Or, soak in lemon juice and salt, then wash.

Or, dab with eucalyptus oil or methylated spirits.

Jam

Soak in a borax solution.

Linen (brown storage spots)

Sponge with bicarb soda moistened with lemon juice. Follow with a normal wash.

Lipstick

Moisten with eucalyptus oil but do not rub. Dab the spot with absorbent material. Repeat until stain disappears.

Mildew

Moisten with a lemon juice and salt paste, and leave overnight. Then wash in warm soapy water, rinse well and dry in the sun. Alternatively, use a vinegar and bicarb soda paste.

Milk

Soak in cold water, then wash normally.

Nappies

Sprinkle bicarb soda between wet nappies in storage bucket to control odour. Seal the bucket with a lid.

Rinse urine-soaked nappies in a vinegar solution (a quarter-cup of white vinegar per bucket of water). Wash in very hot soapy water. (60° C will destroy most harmful bacteria.) Rinse well, adding a quarter-cup of white vinegar to final rinse cycle. Dry in direct sunshine and let nature do the sterilising. Vinegar disinfects, bleaches, deodorises and removes all traces of soap residue from the nappies, thereby ensuring minimal risk of nappy rash.

Remove faeces, then soak nappies overnight in a solution of half a cup of borax (or bicarb soda) to an average machine load of warm water. Wash and rinse thoroughly the following morning, adding a quarter-cup of white vinegar to the final rinse cycle.

To bleach naturally, use any of the following: lemon juice, white vinegar, sunshine or frost.

To soften naturally, use a weak solution of any of the following: vinegar, bicarb soda or Epsom salts.

Perfume

Rub glycerine into the stain and leave for 2 hours. Now sponge with borax and warm water. If the smell still lingers, soak in a vinegar solution, then hang in the sun for several days.

Perspiration

Soak in a warm vinegar or lemon juice solution for 30 minutes. Rinse well, then wash normally.

For wool, dissolve bicarb soda in warm water and soak stain for 1 hour. Follow with a gentle sponge, using warm water with a little vinegar added.

Rust

Moisten salt with lemon juice. Apply paste to rust marks and work into the stain. Leave for 10 minutes, then rub clean. Rinse well with cold water.

Scorch marks

Dampen with water, then dust with borax. Leave to dry, then brush clean.

Or, rub lemon juice into the scorch mark and let it dry in the sun. This will usually remove the mark. A heavy frost restores white crispness as well.

Scorch (on wool)

Rub with a piece of dry steel wool, then sponge with a weak borax solution.

Shoe polish

Dab with eucalyptus oil or methylated spirits.

Socks

Soak children's dirty socks overnight in salty water, then wash normally.

Whiten dirty white socks by soaking them in borax and warm soapy water, then drying in sunshine or frost.

Soot

Cover the soot with a thick layer of salt. Leave for

10 minutes. Lightly sweep the salt and soot into a dustpan.

Tar

Soften tar (or bitumen) with eucalyptus oil, then scrape off as much as you can. Finally, dab clean, using more eucalyptus oil.

For very stubborn tar stains, saturate with eucalyptus oil or glycerine, cover with a bicarb soda and vinegar paste and leave for 1 hour. Now wash in warm soapy water, rinse and dry in the sun.

Tea

Soak or sponge using borax and cold water, then wash normally. Even long-standing stains benefit from this treatment. Glycerine will also soften tea stains.

Urine (wet mattress, pup's puddle or cat's urine)

Soak up the urine with a towel, then sprinkle the mattress with bicarb soda. If possible, let the mattress dry in sunshine. Vacuum away the dry bicarb soda. Repeat this process several times. As a final touch, you may choose to spray lightly with eucalyptus oil.

Place absorbent paper or a towel over the pet's 'puddle' and tread in to maximise urine absorption. Sponge with bicarb soda dissolved in warm soapy water. Give a final rinse using vinegar and-water to eliminate odour.

Sprayed cat urine is particularly offensive and a common problem. Absorb liquid with a paper towel, then coat liberally with dry bicarb soda.

Leave overnight (or for as long as possible) to absorb smell. Then sweep away the bicarb soda and finish with a vinegar wipe-over. Do not use ammonia-based cleaners as they have an old-urine smell to cats that will encourage further urination.

Vegetable

Sponge with a borax solution.

Vomit

Use a large spoon to remove surface material. Sprinkle carpet generously with bicarb soda. Place absorbent paper or towel over the stain and tread in to maximise absorption of liquid. Sponge with warm soapy water. Give a final wipe-over with vinegar to eliminate odour. Use a similar procedure for vomit-soaked clothing or bedding.

Wine (red)

Sprinkle salt and soda water (or bicarb soda solution) over the stain as soon as possible. Blot thoroughly, using absorbent paper, then rinse with cold water and wash in the usual way. If a stain remains, soak in borax and warm soapy water.

Wine (white)

Sprinkle soda water over the stain, then soak up liquid with a towel. Follow with a normal wash.

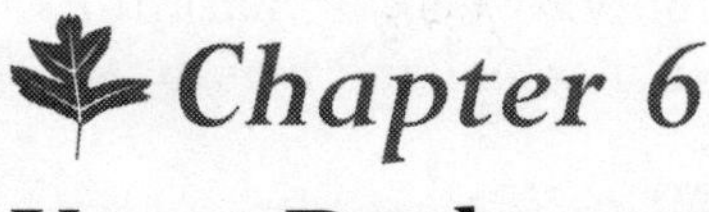

Chapter 6

Your Bathroom

Do you clean your bathroom and end up with streaming, bloodshot eyes, a headache, rash, sneezing or draining fatigue? If you suffer from any of these conditions, the reason could be the cleaning products you're using. There is no need to feel this way.

Why use toxic cleaning products when there are safe green alternatives? Why buy expensive cleaners when simple budget-priced products are readily available? Use the following hints as a guide to direct you towards safe, effective and thrifty green alternatives.

Air freshener

Commercial air fresheners tackle smells the wrong way. It's wiser to deal with the source of the problem and then improve ventilation rather than mask unpleasant smells with artificial fragrances.

Fresh flowers provide a pleasing natural fragrance. Place a bowl of roses, a vase of lavender, a posy of geraniums, sprigs of rosemary or mint, or a bowl of pot pourri on the bathroom shelf. Every time you use the toilet, take a moment to crush a mint leaf and enjoy its clean, sharp fragrance. Eucalyptus oil (or spray) can also be used as a natural air freshener.

Indoor plants have the ability to soak up harmful odours; likewise activated charcoal and bicarb soda.

Bad stains

Rub the stain with a paste made of bicarb soda and vinegar. If the stain persists, leave the paste to sit for a while. Now rub some more, then rinse clean with hot water.

Or, soak the stain in pure lemon juice (for about 30 minutes). This will often soften it enough to rub clean with bicarb soda.

Or, leave a paste of borax and lemon juice on the stain until the stain dissolves.

For really stubborn stains, try an application of eucalyptus oil. Leave for 1 hour, then clean away the stain. Rinse with hot water.

If you've let the bathroom get way out of control, make up a paste using borax, a few drops of your favourite oil (eucalyptus, lavender, citronella, tea tree or peppermint), a squeeze of environment-friendly detergent and some vinegar – then rub clean and rinse.

To give a final shine, wipe over the surface with a cloth moistened with white vinegar.

Bath, shower recess and vanity basin

By wiping over wet areas every day with a dry towel, there will be very little build-up of soap scum, bacteria or mould. A quick weekly vinegar wipe-over will leave the surfaces disinfected, deodorised and shiny, as well as clean of mould and mildew.

Bicarb soda on a moist cloth is ideal for cleaning lightly soiled surfaces.

Combs and brushes

Soak in a vinegar and water solution, followed by a warm soapy wash. Use an old toothbrush to scrub extra-dirty combs. Rinse well and dry in the sun.

Alternatively, dissolve 1 tablespoon of bicarb soda in a basin of very hot water. Soak combs, then clean with an old toothbrush and rinse and dry in the sun.

Grouting

To clean, make up a strong solution of Epsom salts, then scrub with an old toothbrush. Rinse with a hot water and vinegar solution – the vinegar will protect against mould. Or, use equal parts bicarb soda and borax moistened with vinegar, on a nail brush, to cut into the grime. Rinse with a hot water and vinegar solution.

To prevent a build-up of grime between tiles, wipe over regularly with white vinegar.

Mirrors

Clean with a cloth or a ball of newspaper moistened with vinegar.

To remove spots and polish the surface, use a soft cloth moistened with cold tea.

To reduce fogging, use a ball of newspaper moistened with eucalyptus oil.

Mould control

A mould problem can be solved by following these hints:

- Use an exhaust fan in your bathroom.
- Open doors and windows to ensure ample cross-ventilation.
- Wipe over wet areas daily.
- Clear away overhanging trees, creepers and shrubs from around your home.
- Check for leaking taps or pipes.
- Check hidden areas (for instance, on top of blinds and high windowsills) for dust, as dust and mould tend to live together.
- Dry towels, bathmats and face washers daily. Damp towels can develop a musty smell due to mould.

To clean mould from walls and ceilings using green cleaning products, you will need:

- Four containers, clean cotton cloths, bicarb soda, vinegar and hot water.
- Container *one* holds a mixture of bicarb soda and hot water (3 tablespoons bicarb to 2 litres water).
- Container *two* holds a mixture of vinegar and hot water (half a cup of vinegar to 2 litres water).
- Containers *three* and *four* hold hot water only, and are used to rinse the cloths between use. Remember to rinse the bicarb soda and vinegar cloths in different rinsing buckets.

First use the bicarb soda solution, then the vinegar.

If a patch is particularly difficult to remove, bicarb soda or borax can be used dry, as an abrasive.

Eucalyptus oil is useful as an anti-mould and anti-mildew spray.

Rust

Moisten borax with lemon juice, then rub into the stain. Leave for a few minutes before rinsing clean.

Shower curtain (mildew stain)

Wash in hot soapy water, rinse, then 'paint' lemon juice onto the stain and leave to dry in the sun. Alternatively, moisten with lemon juice and salt. Leave for half an hour or so, then rinse well and drip-dry in the sun.

Shower curtain (new)

A new shower curtain is best soaked in salty water before use to prevent the growth of mildew.

Shower curtain (soap build-up)

Soak in a bucket containing 1 cup of vinegar to half a bucket of very hot water. Then machine wash with your normal soap powder and half a cup of bicarb soda, along with a couple of towels. Add 1 cup of vinegar to the rinse water. Hang out in the sun without spin-drying.

Shower rose (to clean)

Unscrew shower rose and soak for an hour or so in white vinegar. Brush rose with an old toothbrush while still in the vinegar solution, then rinse well.

Shower screen (glass)

Clean with a soft cloth moistened with vinegar, or bicarb soda on a damp cloth.

If the soap build-up is difficult to move, mix a paste using bicarb soda, a little eucalyptus oil, some environment-friendly detergent and vinegar. Rub vigorously, then rinse clean.

Taps

Clean with a cloth moistened in vinegar. For stubborn stains on and around taps, rub with a borax and lemon juice paste, rinse well and polish dry.

Taps (stains made by dripping)

Rub with a borax and lemon juice paste. Leave for 5 minutes, then rub clean. Rinse well and dry with a soft cloth. Repeat if necessary.

Tiles (ceramic)

Wipe with a soft cloth moistened with white vinegar.

Or, apply a borax (or bicarb soda) paste to the tiles, then scrub clean with a small brush. Rinse well.

Alternatively, use dry steel wool on dry ceramic tiles to remove soap scum. The soap scum comes away easily as a powder. Avoid breathing in this dust.

For a routine wash of a tiled floor, add 1 cup of vinegar to half a bucket of very hot water, then use your mop to clean and disinfect the floor.

Toilet

The following procedure is safe for both septic and town systems:

- Wipe toilet seat with a cloth moistened with vinegar to give a clean, shiny, odourless surface.
- Add 1 cup of vinegar to the bowl and leave

overnight to soak. Scrub with a toilet brush the following morning.
- Remove any stains by applying a paste of bicarb soda and lemon juice. Rub clean after half an hour or so.
- If the whole bowl is stained, add half a cup of vinegar and half a cup of bicarb soda to the water, then use the toilet brush to scrub the sides. You may need to leave it to soak for a while.

Robin's favourite green hints

Unfortunately I haven't been able to abolish housework, but I have managed to make it quick and easy, and I have replaced all harsh cleaning products with effective green alternatives.

We have an arrangement in our home that the last person to use the shower and hand basin in the morning has the pleasure of wiping it out with a dry towel. Admittedly, that person is nearly always me! However, I really don't mind because it's an easy task when done on a daily basis, and it reduces the amount of soap build-up, as well as lessening the likelihood of mould.

Vinegar is my favourite cleaner for use in the bathroom. Bicarb soda follows. If that doesn't work, then I try vinegar and bicarb soda together. Borax comes next, then borax and lemon juice. Finally, eucalyptus oil and steel wool – but they're rarely needed.

Since the bathroom is necessarily a warm, moist room that is susceptible to the growth of mould, we always use our exhaust fan when showering. When the bathroom is not in use, the door and window are left open to ensure cross-ventilation.

Chapter 7
Your Living Rooms and Bedrooms

The Greenhouse Effect, ozone depletion, air and water pollution, loss of forests, salinity and soil erosion: these are ecological warning signs. Affluence did not have to lead to over-consumption and an escalation of waste and pollution problems. Perhaps now, in the twenty-first century, we have the opportunity to take stock of what we really need, assess living standards in the light of conservation, ecology and health. It isn't difficult; in fact it's all very simple and so inexpensive!

Air-conditioners

Reverse-cycle air-conditioners are a very efficient and economical form of heating and cooling. It's important to clean the filter regularly (at least once a month), as a build-up of dust and mould may have occurred. Brush off the dust, then use bicarb soda to clean away any visible mould. Rinse with hot water and vinegar.

Armchairs (greasy upholstery)

If your armchairs have a slight overall greasiness

or odour, you can dry-clean them yourself. Simply fill a sieve with bicarb soda, then sprinkle the powder through the sieve and onto your upholstery. Using a soft brush, work the powder into the fabric, then leave overnight or for at least half an hour. Finally, vacuum thoroughly and you'll find the bicarb soda has absorbed any odour, grease or loose dirt.

Blinds

Conventional pull-down blinds are easy to clean. Every month or so, wipe along the top of roller blinds with a cloth moistened with white vinegar. The vinegar will prevent the growth of mould and mildew and also clean away any surface dust. Spots on the blind can be sponged using soap and warm water, or vinegar and water. Alternatively, spots can be rubbed away using an eraser.

Venetian blinds (aluminium) respond well to dusting with a feather duster every few weeks or so. To wipe clean, wrap a soft cloth around the blade of a round-ended knife, moisten with vinegar, then rub carefully along each slat. Alternatively, use warm soapy water or methylated spirits. If the Venetian blind is really dirty, you will need 2 cloths, 2 round-ended knives and 2 containers of water – one of warm soapy water for washing, the other of vinegar and hot water for rinsing. Timber Venetian blinds can be cleaned in the same way, but keep your cloths and the wood as dry as possible.

Roll-down split bamboo blinds need a light dusting every now and then. If they are dirty, however, take them outside and clean using a bucket of warm salty water, a soft brush and your garden hose. Select a day that is warm, with a light wind, and take care that the bamboo dries flat.

Vertical blinds need light dusting with a feather duster. Sponge away any dirty spots using a soft cloth moistened with vinegar.

Outside Holland blinds are best swept using a soft broom to remove cobwebs, dust and other insects. Sponge any dirty spots using warm salty water. Rinse with warm water and vinegar. Alternatively, use a soft eraser to rub away any small spots.

Books (leather- or linen-bound)

To polish and condition leather (and old linen) bindings, rub in a little neatsfoot oil, then polish with a soft dry cloth.

Mildew on leather (or old linen) covers can be removed by wiping with vinegar, then allowing to dry in the sun. Repeat if necessary. A little eucalyptus oil on a soft cloth can be used to remove stubborn stains.

Books (musty smell)

Place bicarb soda in a fine sieve, then dust the bicarb soda lightly between the pages. Leave for a few days, then blow away the bicarb soda and the musty smell!

Books (wet)

Place wet books in your freezer until the water crystallises and can be easily brushed off the pages.

Bricks

Bricks around fireplaces can be kept clean and richly coloured by washing with a vinegar and hot water solution, the strength depending upon the amount of grime to be removed.

To add a pleasing finish, mix 1 part lemon juice (strained) to 2 parts olive oil and apply with a damp cloth. Polish lightly with a soft dry cloth.

Camphorwood chest (carved)

To clean away dust and grime trapped in carved areas, use a firm toothbrush dipped in warm soapy water to which you've added a little eucalyptus oil. Dry immediately with a towel, using a hair dryer to dry the deeper crevices. When the chest is clean and dry, use a soft dry toothbrush to work almond oil into each and every surface. Polish the chest with a soft cloth. Finally, to renew the camphor smell, rub the inside of the chest with fine sandpaper.

Cane chairs (sagging)

Tighten up saggy cane chairs by scrubbing them with hot soapy water, then leaving them to dry outside in the wind. The cane will shrink and regain its firmness.

Cane ware

Cane furniture can be cleaned with a soft brush and warm salty water. Dry with a soft cloth and rub over with a little olive oil.

Alternatively, apply a solution of equal parts vinegar and water. Dry in a windy position, then buff with a soft cloth to give extra gloss.

Children (health of)

Young children deserve the very best care you can give, which includes a home where the use of chemicals is kept to a bare minimum.

Grandparents as well as parents need to be aware of this. Good ventilation within the home is essential, so open up those windows and doors.

Coins

Wash coins in very hot soapy water, then soak overnight in lemon juice to remove tarnish. Now wash again in warm soapy water, rinse in hot water and polish dry with a soft cloth.

Combustion heater (to clean glass door)

Using a hot moist cloth, dip in wood ash and rub the glass clean. Give a final wipe-over with vinegar, then polish dry.

Cork mats

Wash in hot soapy water, then rub dirty spots with a pumice. Rinse well and dry carefully. Pumice is spongy, volcanic lava that is useful as a gentle abrasive and polishing material. You can buy it from your pharmacy.

Cowhide

Vacuum or dust the skin. Wash by hand, using pure soap and a little warm water. Rinse clean with a damp cloth moistened with vinegar. Buff the surface with a soft dry cloth.

Cupboards (damp)

Place a few pieces of chalk on the shelves to absorb moisture and prevent the growth of mildew. Spray lightly with eucalyptus oil and improve ventilation.

Curtain rods

Candle wax or soap smeared on curtain rods will encourage the smooth pulling of curtains.

Dent in carpet (caused by furniture)

Vacuum carpet, then dampen a piece of white towelling and squeeze out excess water. Place cloth over flattened pile. Iron for 15 seconds (with iron set on 'high dry') to allow steam to lift the carpet pile. Remove cloth and allow carpet to air dry.

Deodorise (a room, wardrobe, drawer or car interior)

Stale cigarette smoke, perfume, sweat, vomit, spilled milk, wet dogs, lingering food odours like garlic or fish and chips, and mildew on carpet – all these and more cause problems in homes and in car interiors.

Fresh air is both free and the very best deodorant, so open up those doors and windows! However, if nature needs a little help, use any of the following in a small bowl: vinegar; herbal vinegar; cloves; activated charcoal; fresh herbs or sweet-smelling flowers; or any of the fragrant oils, such as eucalyptus, tea tree or citronella.

Or, boil a handful of cloves in a saucepan of water for a few minutes. Alternatively, place an open packet or bowl of dry bicarb soda in the area to absorb smells for about 3 months.

Careful selection of building materials and furnishings, in relation to their emission of harmful fumes, will ensure healthy, people-friendly indoor environments, especially when good ventilation allows free air-flow.

Drawer (that is sticking)

Rub soap or candle wax over the tracks to give a smooth, easy-running drawer.

Finger marks (on painted surfaces)

Use vinegar on a moist cloth.

Or, clean grimy patches using bicarb soda on a damp cloth. Bicarb soda will not scratch the surface.

Or, wash with a hot soap and borax solution.

Floor rugs (cotton, wool or synthetic)

To clean, first vacuum, then shake away all surface dust. Now wash using pure soap, borax and warm water. Rinse in a weak vinegar solution, then dry in the sun, preferably on a windy day.

For musty rugs and carpet, sprinkle with plenty of bicarb soda, leave overnight, then vacuum the following morning. Hang rugs out in the sunshine.

Glass, glass-topped tables and crystal (to clean and polish)

Wash with a solution of 1 part vinegar to 3 parts hot water. Do not rinse. Crystal is best air-dried. The vinegar ensures a clean, shiny and streak-free surface.

Lampshades

The delicate fabric of an old lampshade can be safely cleaned using dry bicarb soda or powdered borax. Rub the powder carefully into the material, then dust lightly for a pleasing finish.

Leather

Leather breathes well, cleans easily, has a soft luxurious feel and looks and smells great. To clean leather, rub dirty marks with a moist soapy cloth, then wipe clean with a second moist cloth. Polish dry immediately using a third cloth.

Leather clothing needs to be stored in an airy cupboard.

Any of the following preparations is suitable for revitalising leather surfaces. Each is applied with a soft cloth, then buffed dry to give a pleasing sheen:

- Equal parts olive oil and vinegar.
- Vinegar (this will discourage the growth of mould in areas of high humidity, especially on leather goods stored in cupboards).
- One part vinegar to 2 parts linseed oil (this will help prevent cracking).
- A little castor oil (for dark-coloured leather).
- A little light olive oil (for light-coloured leather).

Leather boots (to waterproof)

Mix together equal parts beeswax (warm in a double saucepan) and castor oil, then rub the mixture into the leather while still warm.

Mould and mildew

Black, green-black or brownish specks or stains on your walls, ceilings and in the grouting between tiles are minute furry growths of fungi – mould and mildew. These thrive in damp, warm, poorly ventilated houses. They can also live under carpets and in cupboards. In hot, humid climates, verandah rails, toothbrushes, shoes, handbags and clothes stored in poorly ventilated cupboards frequently grow mould and mildew.

Here are some hints to minimise mould and mildew:

- Always dry shoes, bags, coats, umbrellas and luggage before putting them away in a well-ventilated rather than airtight cupboard.

- Store linen and clothing in airy cupboards.
- Wipe over any mould-susceptible surface with vinegar.
- Rinse any mould-susceptible fabric in a vinegar and water solution.
- Choose the sunniest room for young children to sleep in, with good through ventilation.
- Give mattresses, underblankets and pillows a regular 3 hours in sunshine.
- Wipe over the leaves of indoor plants regularly using a soapy cloth. Also, remove dead or diseased leaves. Turn over the soil every week or so.
- Clean around the vents of your air conditioner; let cleaning reach to the tops of blinds and high windowsills; avoid cellars and antique furniture; wipe around window frames and keep aquariums clean.

To kill and clean away mould and mildew, use a strong solution of any of the following: vinegar, lemon juice, salt, Epsom salts, eucalyptus oil and tea tree oil (with the oils, add a few drops to a basin of very hot soapy water). Simply wipe over the surface, leave for 10 minutes or overnight, then finish with a vinegar and hot water rinse.

If the mould is particularly ingrained, combine bicarb soda (or borax) with vinegar and use a stiff brush to clean the area.

Or, moisten with lemon juice and salt. Leave for 30 minutes, then rinse well and, if possible, dry in the sun.

The good news is that these green cleaners really work.

Paint ('Breathe Easy' Flat, Low Sheen, Semi Gloss and Sealer Undercoat; all acrylic, for interior use only)

No solvent fumes (and very little odour) are released into the environment during or after using this water-based paint. For young children and those adults who react to paint fumes, this may be a safe option. Asthma and other respiratory reactions can be triggered by conventional enamel and acrylic paints. It's a good idea to paint a small area, then judge the results for yourself. You can buy small test pots of paint from good hardware and paint stores.

Paint (cleaning up leftover)

Leftover paint should not be poured down drains or onto your garden. Unwanted paint is best brushed out onto newspaper, allowed to dry in an airy shed and disposed of according to local council regulations. Empty paint containers should be left open in a well-ventilated outside area to dry out before disposing of them.

Painted surfaces (to clean prior to painting)

Wet the wall with plain water using an old paint brush. Wash the surface using a 'Lectric' soda solution, then rinse with clean water. Use a soft cloth to wipe dry. Smoke and grease stains require a strong solution. Alternatively, use a hot soapy water and borax solution.

Paint (fumes from)

When painting, regardless of the type of paint used, it's wise to have all your windows and doors wide open to allow good through ventilation. It's also a good idea to avoid sleeping in freshly painted rooms for at least 1 week, and for pregnant women and young couples wishing to start a family to avoid paint fumes altogether.

Paint spill

Soak up a spill immediately using dry sand or cat litter (white clay particle type). Now scoop up the mixture and place it in a labelled container. Check with your council for methods of safe disposal. Make sure it doesn't get into drains or waterways.

Patent leather

Rub with the inside of a banana skin and leave to dry, then polish with a soft cloth until gleaming.

Or, dab on a light application of milk with a piece of cotton cloth, then buff with a dry cloth to produce an excellent surface.

Pictures (glass-covered)

Clean the glass using a little methylated spirits on a soft cloth. Never use water, as water may seep beneath the glass and ruin the picture.

Plaster

Adding a little vinegar to the plaster mixture will delay the hardening.

Plastic toys

If a child sucks or chews soft plastic toys, there's a danger that the plastic will be swallowed. It's wise to restrict plastic toys to those made of hardened plastic only. Lots of great toys are made of natural products such as wood, metal, cotton and wool. But be aware that glues, paint, varnish and rubber, which are best avoided, may have been used in their production.

Polished timber floors (and pre-sealed panel-board flooring)

Your vacuum cleaner is an excellent way to pick up dust and all sorts of other debris that collects on floors. Alternatively, use a broom, a dry mop or a mop moistened with cold tea or vinegar.

To clean, use 1 cup of white vinegar to half a bucket of very hot water. The vinegar will clean, disinfect, deodorise, act as an anti-mould agent, prevent any 'spotting' as the floor dries and leave a shiny surface. Be careful not to over-wet the floor.

For scuff marks or any other stubborn stains, use a little eucalyptus oil or methylated spirits.

Porcelain

Rub stains with lemon juice, rinse clean and polish dry with a soft cloth.

Posters

An old poster can be removed from a wall by painting the poster with a few coats of vinegar, allowing the vinegar to soak through the paper, then sponging with warm water and vinegar and peeling it off.

Price tags and labels

Moisten adhesive labels with eucalyptus oil, then peel free. To remove sticky adhesive residue, simply rub lightly using a cloth moistened with eucalyptus oil.

Reading glasses and sunglasses

A drop of white vinegar on each lens followed by a quick rub with a soft cloth ensures a clean, streak-free surface.

Records (cleaning vinyl)

Using a soft, clean, lint-free cloth (or soft brush) and wiping in a circular motion with the grooves, clean the vinyl with pure soap and warm water. Rinse with lukewarm water to which you've added a little vinegar. The vinegar will remove all soap residue and leave a streak-free, shiny surface. Dry with a soft clean towel. It's surprising how often the simplest things work best!

Stickers, labels and tape can be loosened and sticker residue removed with a little eucalyptus oil or citrus-based cleaner.

If you have a valuable collection of vinyl records in need of cleaning, it's wise to test out your cleaning procedure on a cheap, easily replaceable record. Most opportunity shops sell records inexpensive enough to experiment on.

Seagrass matting

Never use soap on seagrass matting. Simply vacuum or brush clean, then wipe over with a salt and water solution. The salt prevents darkening of the seagrass.

Scissors (to sharpen)

Use scissors to cut thin strips of fine sandpaper or 'wet and dry' paper (available at hardware stores), or fine steel wool into small pieces.

Sheepskin rug

Rub vinegar into the back of the skin before washing the fleece with warm soapy water and borax. Add vinegar to the rinse water then spin-dry, shake and dry in the wind. Rub the back of the skin to keep it supple.

Suede (shoes, bags, clothing)

Suede breathes easily and is softly flexible, making it pleasant to wear and use. Use any of these hints to clean your suede and give it a lift:

- Rub with a soft pencil eraser.
- Brush vigorously using a rolled-up pair of old pantyhose.
- Rub gently with steel wool.
- Boil a kettle and hold the article over the steam for a minute or so, then brush with a fine suede brush to rejuvenate the flattened suede.
- Sprinkle with cornflour, then brush clean after 5 minutes.

Suede (to remove nail polish from)

Nail polish remover is not recommended, for two reasons. Firstly, since suede is porous, nail polish remover will spread the polish further into the suede. Secondly, nail polish remover usually contains acetone, a flammable solvent that is toxic when inhaled. Instead, try the following: using a fine suede brush or steel wool, gently work away at the polish. It'll take time and patience, but it won't spread the stain. Try to avoid brushing beyond the stain.

If all else fails, cover the stain with a decorative buckle or braid.

Telephone and mobile phone

A regular wipe-over with a little white vinegar on a cloth cleans and disinfects your telephone.

Television and computer screens

Moisten a soft cloth with vinegar, then clean and polish dry the screen.

Timber furniture (unpainted)

Alcohol spill

If alcohol has been spilled on a polished wood surface, soak up the liquid immediately with a towel, then rub olive oil into the surface grain. Leave for a few minutes, then polish with a soft dry cloth for a pleasing finish.

Bruised furniture

Sponge depressions with warm water. Soak brown paper in water, fold over several times and place over the bruised wood. Use a hot iron to evaporate the moisture from the paper. Repeat until the wood has regained its structure.

Candle wax

Place a small plastic bag full of ice over the candle wax for 5 minutes. Scrape off the hardened wax, then clean with a little eucalyptus oil on a soft cloth. Now polish with equal parts olive oil and vinegar.

Clean and polish

Any of the following preparations are suitable for revitalising timber surfaces. They are all applied with a soft cloth, left for 5 minutes to dry, then polished with another cloth to give a pleasing sheen.

- Equal parts olive oil and vinegar.
- One part lemon juice (strained) to 2 parts olive oil.
- Equal parts linseed oil and vinegar, especially for dark wood.
- Full-strength vinegar on a moist cloth, to clean and discourage the growth of mould.

Cracks

In dry places where the humidity is low, it's a good idea to place a small bowl of water beneath any piece of treasured timber furniture to prevent the timber drying out, which may lead to cracking. Cracks, which can also be caused by direct sunshine, can be mended by melting beeswax to the consistency of putty and smoothing it into the cracks. Now lightly sandpaper the surrounding timber and blend the wood dust into the beeswax. Allow the beeswax to set, then polish.

Dark stains

Bleach dark stains using white vinegar.

Or, moisten salt with lemon juice and rub into the stain. Leave to dry, then brush off.

Grease spots

Mix equal parts chalk and bicarb soda with water to make a paste. Spread over grease. Brush off when dry.

Nail polish and paint splashes

If spilled on polished wood, mix equal parts vinegar and olive oil and apply to the nail polish using a warm moist cloth. Rub until clean.

Alternatively, use a salt and olive oil paste.

Polish build-up

To remove a build-up of polish on furniture, sponge with a cloth moistened with vinegar. Allow to dry, then re-polish.

Raw wood (to stain and condition)

Rub in equal parts oil (olive, linseed or any other good quality vegetable oil) and vinegar. Repeat several times.

Stain to your preferred colour using natural pigments such as sienna, iron oxide etc. Then rub in a couple of applications of oil.

Or, make up a mixture of 1 part beeswax (heated) with 4 parts linseed (or olive) oil. Rub into the wood, leave 30 minutes, then polish with a soft cloth. You can add a few drops of fragrant oil to this mixture, if desired.

Scratched furniture
Has a boisterous pup or kitten scratched your furniture? Find a crayon of the same colour and fill the scratch with the crayon wax.

Or, match the scratched wood with pecan or walnut 'flesh', then rub the flesh into the scratch until the mark disappears.

Finish by rubbing with a 2 parts olive oil, 1 part vinegar mixture, working along the grain. Polish dry with a soft cloth.

White marks
Ugly white marks (caused by placing hot cups on a polished surface) can be removed in the following way. Add pumice powder (or fine wood ash) to olive oil (or raw linseed oil) to make a thin paste. Rub into the damaged area following the natural grain. Now moisten a cloth with plain olive oil (or linseed oil) and wipe off surface grit. Repeat process until the mark disappears.

Varnish (to remove old varnish)

Scrub with a stiff brush using a borax and hot water solution.

Wallpaper

Crayon marks
Place a piece of soap within a pad of steel wool and rub gently.

Grease spots
Rub a thick piece of stale bread over the spot until the grease is absorbed.

Or, mix equal parts chalk and bicarb soda with

water to make a paste. Spread over grease, leave until dry, then brush clean.

Removing wallpaper
Mix equal parts vinegar and warm water. Sponge wallpaper until thoroughly wet. Peel off.

Smoke stains
Sponge with bicarb soda and warm water.

Sticky tape on wallpaper
Press with a warm iron.

Vinyl wallpaper

Wipe clean with a vinegar and hot water solution.

White paint (discolouration)

Wash with a solution of borax and hot soapy water. Finish with a vinegar and hot water rinse.

WINDOWS AND FLY-WIRE SCREENS

Fly-wire screens

If possible, take your fly-wire screens outside to brush away loose dust, spider webs and insects. Make sure that the wind carries the dust away from you. Using vinegar and very hot water (1 cup of vinegar to half a bucket of very hot water), wash screens and clean away any traces of mould and mildew. Give a final wipe-over with pure vinegar, or eucalyptus or tea tree oil (a few drops in a basin of very hot water). Finally, dry the screens in the sun.

For screens that can't be taken down, use a vacuum cleaner to suck away dust, sticky spider webs, insects and mould spores. Then wash and wipe over, as above, and allow to dry.

Mould is often a problem on south-facing screens,

and a real issue for people with asthma and other allergies. It's very important, in terms of the quality of air you breathe indoors, that your screens are clean of dust and mould spores.

Glass (including car windows and windscreens)

Add 1 cup of vinegar to half a bucket of hot water. This washing solution will help prevent 'spotting' on the glass while cleaning at the same time. To avoid ugly smears, clean the outside with vertical strokes and the inside with horizontal strokes (or the other way around). This way it's easy to see and wipe away smears. If windows are very dirty, use hot water and detergent.

For dirty marks and smudges left by children's hands and dogs' noses, rub clean using a dry lint-free cloth. This is the easier method!

Or, moisten a cloth with vinegar, then rub gently. Polish dry with another clean cloth.

If stickers have left adhesive material behind, apply eucalyptus oil, then rub clean.

Windowsills

Use a soft cloth moistened in a warm vinegar solution to clean windowsills and remove any mould.

INDOOR PLANTS

Research has shown that indoor plants are efficient air purifiers. They also help to create an atmosphere of peace and relaxation. In return, indoor plants require adequate levels of light, moisture, nutrients, warmth and humidity in order to thrive. They also need to be kept free of dust and away from fires, radiators and draughts.

Dust on leaves

It's very difficult for plants to 'breathe' if their leaves are covered with dust. Weekly 'misting' with water (using a plant sprayer) helps remove dust, and so too does a simple wipe-over of the leaves using a moist cloth.

Pests on stems and leaves

Aphids, mites, scale and sooty mould can easily be controlled by wiping the stems and leaves with a soft cloth moistened with soapy water. Aphids can be squashed between your fingertips, and scale insects removed manually using a stiff toothbrush or by scratching them off using your fingernails. Or, use cotton buds dipped in methylated spirits to kill any small insect pest.

Alternatively, combine some water with a little vegetable oil and a drop or two of detergent. Shake up the solution, then either spray or wipe over the leaves. This suffocates plant pests. Other safe sprays include chamomile, milk and bicarb soda. It's as simple and inexpensive as that! See pages 116–125 for more information on organic sprays.

Any plant under stress due to lack of light or moisture will be more susceptible to damage by pests.

Soil (care of)

Mould and mildew can establish themselves on the surface of soil, so it's a good idea to turn soil over on a regular basis, using an old kitchen fork. Over-watering results in root rot and the growth of mould and mildew.

Terracotta tiles (stained with water from indoor plants)

Flood with water to which a little vinegar has been added. Now sprinkle with cat litter (white clay particle type) to absorb the 'salts'.

Robin's favourite green hints

A friend's cat recently came in through her cat flap with a smelly rat. After disposing of the rat, my friend emptied two packets of bicarb soda over the carpet to cover the smelly trail. She left the bicarb soda to work overnight, then vacuumed it up the following day. All the smell had gone.

A lint-free cloth moistened with vinegar is my way of cleaning our telephone, computer keyboard, mouse, printer, TV and video, and car steering wheel.

Living so close to the ocean means that our windows are often coated with salt spray – as well as dust – in the drier months. We find that a quick wash-over with hot water and vinegar works wonders.

My favourite pieces of furniture include a small silky oak table that my father made and an English oak desk that was my grandfather's. These pieces of furniture are polished using equal parts olive oil and vinegar. This way I preserve their rich colours and attractive grain and keep them looking clean, polished and natural. They should last forever.

To provide a protective coating for untreated wood, we have successfully used olive oil, rubbing in several coats. The result is a pleasing natural finish with no toxic fumes.

Chapter 8
Indoor Pests

Green pest control relies on a basic understanding of the creature involved and the wise use of barriers, traps, baits and repellents, as well as high standards of hygiene around the home.

By making good use of a fly-swat, vacuum cleaner, fly-screens, sunshine, mosquito nets, broom, washing machine, iron and freezer, you will successfully practise green pest control. When you treat pests yourself, you decide what method or product is used and how much.

Your first step is a thorough clean-up. The aim is to make pests unwelcome in your home by cleaning away all food and water sources and by disturbing breeding and resting places. Food scraps, dust, areas of moisture and dirty surfaces all encourage household pests and must be eliminated

Make sure that all food is stored in sealed containers and that 'wet' areas are dry. Cockroaches and other pests need both food and water!

In summary, modern pest control is largely a matter of:

- Limiting the pest's food and water supply by

setting high standards of hygiene, food storage and rubbish disposal.
- Preventing its entry by using fly-screens on windows and doors.
- Using repellents, baits and traps to deal with the occasional unwelcome guest.
- And being realistic. Aim to control rather than to eliminate, for to eliminate completely is usually impossible.

It's worth remembering that insects could readily survive without humankind, but humans could not survive without insects.

Ants

What's the attraction? Maybe a clean-up is all that's required? Perhaps, by placing your pet's feeding bowl in a saucer of water, or the jam in your refrigerator, the problem will disappear.

If this approach does not solve the problem, trace ant trail back to entrance point and seal access.

Wipe over the trail using strong herbal vinegar to deter the ants' progress. Alternatively, spray the area with eucalyptus oil.

Mix together equal parts icing sugar and borax and sprinkle along pathways and around nest entrances. Remember that borax is a poison.

Or, mix together equal parts jam and borax and place the bait on small lids in strategic places. If your ants prefer icing sugar or cake, use these instead of jam.

If you have to destroy a nest, flood it repeatedly. Alternatively, blend together half a cup of lime with a generous amount of detergent in a bucket, then fill with very hot water. Now pour carefully down the nest's entrance.

Remember, ants are not necessarily pests. They fulfil many useful functions, such as aerating the soil, recycling dead plant and animal material and eating termites, cockroaches, caterpillars, fleas and flies.

Borers

Pin-sized holes and fine sawdust signal borer damage. Select timber resistant to borers, such as cypress pine, western red cedar, mountain ash or redwood.

Strong sunlight, as well as heat, destroys borer larvae. Seal the timber in black plastic and leave in full sunshine for 3 hours. Alternatively, use natural pyrethrum spray.

Carpet beetles

Remembering that it's the beetle that lays the eggs, but the larvae that cause the damage, use the following hints to keep adult carpet beetles out of your home:

- Use fly-wire screens on windows and doors.
- Vacuum regularly, especially under windows, around the edges of carpet and in corners. Dead flies attract adult carpet beetles.
- Check cut flowers for adult beetle activity prior to arranging them.
- Store all flour and cereal products in sealed containers.

To deal with the larvae, shake rugs outside, then hang in sunshine for 3 hours. Vacuum carpet thoroughly, then spray with pyrethrum.

Cockroaches

These oval, flat-bodied insects are experts in squeezing through very small spaces.

Cockroaches require darkness and warmth, and will eat almost anything. Here are some methods to reduce their impact:

- Seal all entry points (holes and cracks in floors and walls, and gaps under doors) to your home. Block chimneys when not in use.
- If possible, install door and window screens.
- Disturb known hideouts as often as possible. For example: warm, dark, moist places beneath water heaters, refrigerators, sinks, drains, wood piles etc.
- Fix leaking taps.
- Keep kitchen benches clean, food in sealed containers, rubbish in a well-sealed bin, and floors swept and clean – especially overnight.

Repellents, baits, traps and sprays

Repellents include Epsom salts sprinkled about, or eucalyptus oil sprayed or smeared around crevices.

Baits, which are simple and very effective, include borax (or alum or natural pyrethrum powder) sprinkled in cracks, crevices, corners and dark places where cockroaches hide, such as behind refrigerators and stoves. The fine powder will stick to the cockroaches' feet and act as a poison when they lick themselves clean. It will take a few weeks for the cockroaches to die.

An exceptionally effective bait is made with equal parts borax and flour mixed with sufficient molasses to form a dough. Push pieces of the dough into crevices, under sinks and anywhere else you see evidence of cockroaches. This dough is long-lasting and effective. Remember that borax can be toxic, so ensure that children and pets are kept well away.

Traps are suitable for small infestations and may be quite simple. For instance, place a piece of dripping inside a plastic container and put it in a dark place near an entrance point. The cockroaches will climb in for a tasty meal, get slippery feet and be unable to get out.

Or, smear vegetable oil around the inside of a large glass jar, then drop in a piece of banana. The banana will attract the cockroaches but they will be unable to climb out of the jar.

Sprays can be useful, especially natural pyrethrum sprayed around cracks and crevices.

Fleas (from pets)

Many people treat their dog or cat for fleas but forget to treat the major source of the fleas: the environment, both inside and out. Did you know that 99 per cent of fleas are found *off* your pet? The natural green approach aims to break the lifecycle of the flea.

Outside, you will need to:

- Block access beneath your house so that your pets cannot rest there and make it an ongoing source of reinfestation.
- Mow, sweep and hose. Kennel areas, doormats, verandahs and other dusty places where pets lie need to be swept, then hosed, to bring adult fleas and larvae to the surface. If the area is grassed, it's a good idea to mow, then rake the grass, prior to hosing.
- Dust or spray. After the hose-down, your next task is to sprinkle the area with lime, natural pyrethrum powder or salt, or to use a herbal spray, in order to kill all adult fleas and larvae. Cover fish ponds and caged birds prior to dusting or spraying.

- Pack salt into cracks in and around kennels.
- Whenever pruning your herbs, collect the cuttings and place them under your dog's bedding and anywhere else your dog lies. Fresh brown pine needles also deter fleas, so they make quite good bedding for dogs. Do check, however, that the pine needles are not mouldy.

Inside, you will need to:

- Vacuum. Fleas, flea eggs and larvae may be lurking in carpets. A thorough vacuum is needed at least once a week of all carpet, rugs and soft furnishings, dusty cracks and corners. Remove the vacuum cleaner bag, then seal tightly in a plastic bag before disposing of it in a responsible manner. Eighty per cent of flea eggs can be removed by vigorous vacuuming.
- Mop. Mop hard floor surfaces using a solution made up of 1 cup of white vinegar to half a bucket of very hot water.
- Wash. Pet bedding needs to be washed on a weekly basis, using eucalyptus oil in the rinse water. Whenever possible, dry bedding in the sun for 3 hours. This will kill eggs and larvae.
- If car travel is a privilege your dog enjoys, vacuum the car seat regularly and wash any car bedding.
- Give infested material the sunshine treatment. Cushions, rugs and curtains can be placed outside in the sun, as 3 hours of sunlight will kill flea eggs and larvae.
- Dust or spray. With carpets, and if the infestation is heavy, use a low-toxicity herbal, natural pyrethrum or wormwood spray or powder. See pages 121–2 and 124 for these recipes. Repeat every week for a month to break the cycle. Wormwood spray has bitter pungent qualities that make it very effective against fleas.

Note: Do not spray or dust near caged birds or fish. Wear protective clothing when applying the spray. Store out of reach of children and pets.

To treat the animal itself:

- Clip any long-haired dog, then groom it, first of all using a brush, then a flea comb. Drown the fleas in a basin of very hot soapy water.
- Add a natural pyrethrum or herbal shampoo to warm water, blend in a few drops of eucalyptus oil, then wash your pet, usually twice a week if fleas are bad.

You will need to treat all the dogs and cats in your household at the one time. Remember that preparations suitable for dogs are not always safe for cats. Likewise, be aware that products appropriate for adult cats are not always safe for kittens.

Green flea preparations may be purchased through pet shops, some health food outlets and selected supermarkets. They usually combine various concoctions of natural herbs and oils.

Flies

Before you reach for that pest strip or spray can and fill your home with toxic vapours, ask yourself two questions. Firstly, why are there flies in your kitchen? What's the attraction? Is it simply a matter of cleaning up and putting away, and fixing that hole in the fly-screen door? Secondly, are there natural repellents, baits, traps and sprays that you could use?

There is always a green solution. For example, if there's a fly in your bedroom that's buzzing around and driving you crazy, don't reach for the spray and drown that single fly in pesticide.

Simply turn off the light and open the door – with a light on elsewhere in the house – and the fly will depart. Alternatively, use a fly-swat or your vacuum cleaner. Now you will be able to sleep peacefully, breathing air that is fresh as well as completely free of flies.

Here are some simple green ways to reduce the impact of flies:

- Reduce their breeding ground by burying all pet droppings, sealing rubbish bins and turning over and adding fresh grass clippings to slow-heating compost heaps.
- Use fly-screens and fly-wire doors to prevent flies coming indoors.
- Use a fly-swat to kill the occasional fly that beats the barriers.
- Do not leave uncovered food lying on benches or your sink, and clean up spilled food.
- When flies are particularly bad, use eucalyptus oil spray as a repellent on door and window frames. Vinegar wiped over windows and mirrors also helps to repel flies.
- Place stems of rosemary on your barbecue. Mint also smells unpleasant to flies.
- Install an outdoors fly-trap that is designed to let flies in but prevents them from leaving.

Head lice

Head lice is a growing problem for child care centres and schools throughout the world. Although many pharmacy-line head lice preparations are available, these sometimes prove too severe for a child's sensitive skin, especially the skin of the head. Also, there's evidence that head lice are becoming increasingly resistant to commonly used lotions. Many simply don't work anymore.

Head lice eggs are about the size of a pinhead, and whitish-grey in colour. You are more likely to find them at the back of the neck and around the ears. The lice are also tiny but can be seen with the naked eye. They are whitish-brown in colour and move very fast.

Apply any of the alternatives listed below, slightly warmed, to dry hair. Massage well, then comb the solution through the hair, using a fine-toothed steel lice comb to remove the suffocated lice. To see what you've caught, wipe the comb on a white tissue. Head lice show up as dark spots on the tissue, while eggs look smooth and pearly. To dislodge tough eggs, you may need to reverse the combing direction. Leave for 10 minutes, comb once more, then shampoo and rinse as normal. Following this procedure each week is a very easy, effective and non-toxic way of controlling head lice, especially when continued reinfestation is a problem.

Effective products, ranging from very simple to more complex, include:

- Ordinary hair conditioner.
- Mineral oil (a non-prescription pharmacy item).
- Apple cider vinegar.
- One part tea tree oil blended with 3 parts olive oil.
- One part neem seed oil blended with 3 parts olive oil, with 2 drops of eucalyptus oil added.
- 5 ml tea tree oil blended with 5 ml eucalyptus oil, with 50 ml of olive oil added.
- Half a teaspoon each of rosemary, eucalyptus and pennyroyal oils blended with 2 1/2 tablespoons of olive oil.

To help remove the firmly attached eggs from the hair, rinse with 1 part vinegar to 3 parts water.

Then use a fine-toothed steel lice comb to remove the eggs.

Wash clothing and bedding (as well as hats, towels and sofa covers) in very hot water. Dry in the sun, then iron.

To deter head lice, try a few drops of manuka oil (a special type of tea tree oil) combed through the hair.

To suffocate nits in eyebrows, rub in mineral oil and leave overnight.

House dust mites

What we call harmless house dust is, in reality, a cocktail of microscopic dust mites (dead and live) and their droppings, along with human and animal skin, hair, bacteria, moulds, plant and food remnants, and lint from furnishings and clothing.

Dust mites serve a useful purpose by cleaning up micro-matter, including shed human skin particles. Therefore, you should aim to reduce the dust mite population rather than eliminate it completely. It's the droppings of the dust mite that cause the allergic reaction rather than the mite itself.

When anyone in the household suffers allergic reactions, it's wise to reduce the level of house dust mites. Through the sensible use of hot water, sunshine, a freezer and an iron, this can be accomplished the green way. Eczema, asthma and other respiratory conditions usually reduce as the dust mite population decreases. A management plan may include:

- Avoiding the use of carpets, drapes and cushions in bedrooms. Vertical blinds are good.

- Daily sweeping or vacuuming.
- Wiping over the floor and other washable surfaces using a cloth moistened with vinegar or cold tea. Mites cannot tolerate acid!
- Cleaning fly-wire screens with vinegar and hot water. Follow this by a wipe-over using pure vinegar, or eucalyptus or tea tree oil (a few drops in a basin of very hot water), then dry the screens in the sun if possible. This is especially important in bedrooms.
- Opening windows to allow free air circulation.
- Washing bed linen weekly using hot water. Water over 55° C spells death to dust mites. Or, add 1 teaspoon of eucalyptus oil to each load.
- Vacuuming mattresses.
- Regularly airing all bedding, pillows and floor rugs in the sun. Three hours of sunshine proves fatal to dust mites.
- Avoiding air conditioning and central heating, as dust mites thrive in uniformly warm conditions (50 per cent humidity is their ideal).
- Washing pet bedding every week.
- Placing soft toys in plastic bags and putting them in your freezer overnight to kill mites lurking in them.
- Ironing to kill house dust mites. With pillow cases, this is especially important due to their close contact with your face.

Mice and rats

Making your home unattractive to rodents is the green way of dealing with a mouse or rat problem. Here are some hints to consider:

- Don't leave nesting material around, such as boxes of rags.

- Keep your rubbish bin well sealed.
- Never leave out dog or cat food, or caged bird seed, especially overnight.
- Deter with mint, quassia chip spray or eucalyptus oil.
- Clean up areas of long grass around your home and along fence lines. Tidy up around garden sheds. Try to remove anything that would provide sheltered resting or breeding places.
- Block holes with steel wool or crumpled aluminium foil, as rodents cannot chew through it.
- Turn over and add grass clippings to slow-heating compost.

Set traps at right angles along a wall, with the trigger of the trap closest to the wall. Tempting baits include chocolate-coated peanuts, raisins, lightly cooked bacon or a peanut butter and honey mix.

A novel bait that kills by constipation rather than poison is made by mixing together either equal parts flour and plaster of Paris with milk; or equal parts icing sugar and cement dust with milk. Mix to the consistency of a dough, then place baits in the ceiling. The bodies of rodents killed this way will not harm any dog, cat or bird that eats them.

Millipedes

If these small, dark, worm-like creatures (with their many, many pairs of legs) invade your home, adopt the green solution. Instead of using a spray, simply use a broom to sweep them into a dustpan, then endeavour to prevent the entry of more by improving the seals around your doorways. Although a nuisance, millipedes are usually harmless and tend to curl up into a ball when alarmed.

Mosquitoes

Reduce their breeding grounds by checking that water does not lie stagnant in roof guttering and pets' water bowls. Pour a film of vegetable oil on the surface of water tanks, keep ponds stocked with frogs or fish, and clean birdbaths on a regular basis.

Use fly-screens and fly-wire doors to keep mosquitoes out of doors. Use a fly-swat to kill the occasional mosquito.

If mosquitoes are a problem at night, suck them off your walls and ceilings with a vacuum cleaner rather than use a toxic spray. And use a mosquito net over your bed. If there are mosquitoes about, babies and young children need to be covered with mosquito netting while resting.

Wear long-sleeved, light-coloured clothing, especially at dusk. Perfumed deodorants and aftershaves often attract mosquitoes, flies and wasps. If these flying insects seem drawn to you, think about trying fragrance-free products.

As a repellent, use citronella, tea tree or eucalyptus oil, and rub on wrists, ankles, neck or forehead. Avoid contact with your eyes and, to avoid skin irritation, use nothing stronger than a 25 per cent concentration of the oil. Dilute these stronger smelling oils with olive or grapeseed oil. This repellent also works when sandflies and ticks are a problem. Alternatively, use a natural pyrethrum-based repellent, or apple cider or herbal vinegar, which is effective when dry on the skin.

Many herbs are useful repellents. If you're out in the garden and are being eaten alive by mosquitoes or sandflies, go to your herb garden, pick a bunch of mint, lavender or any other fragrant herb, then strip off the flowers and leaves. Rub them between your fingers and palms and spread the scent to your ankles, wrists, neck or any other skin you wish to protect.

Plants that release citronella oil into the air, such as citronella-scented geraniums, are particularly good for planting around patios, barbecue areas and doorways.

Moths (clothes)

The crux of green moth control is to create conditions that female moths find so unattractive that they go elsewhere to lay their eggs – conditions like clean materials or materials with an aromatic scent that overrides the human smell.

Although moths lay the eggs, it is larvae that do the damage. They feast on woollens in particular, although they will also munch on perspiration-stained clothing made of other fabrics.

Moth eggs and larvae are destroyed by washing clothes or by hanging them out in the sunshine for 3 hours. A little eucalyptus oil in the rinse water also helps repel troublesome moths. Woollen rugs should be vacuumed and put out in the sun.

Following a wash, store material in linen or plastic bags, or in a cupboard or drawer scented with aromatic herbs or spices. Moths cannot tolerate cloves, and as cloves are cheap, non-toxic and sweet-smelling, they are ideal to place in small muslin sachets and hang in wardrobes, linen cupboards and drawers. Alternatively, simply scatter them.

Moths also dislike the smell of Epsom salts, lavender, sage, wormwood and rosemary, as well as eucalyptus oil. To keep these natural scents strong, replace the sachets every 6 months or so. In this way, you can use specially selected flowers, leaves, herbs and spices to keep your clothes free of moth larvae attack the natural way. A block of pure soap, placed in a cupboard, drawer or wardrobe, acts as an insect repellent due to the oil of citronella or other aromatic oil usually found in the soap.

A final hint: after discarding toxic mothballs or naphthalene flakes, the unpleasant smell can be reduced by sponging drawers and wardrobes with a mixture of equal parts vinegar and hot water, or equal parts lemon juice and hot water. In addition, a few drops of eucalyptus oil, or any other fragrant oil, may be sprinkled or sprayed around the affected area.

Silverfish

Silverfish have flat, elongated, silvery bodies and three tail-like structures. They move rapidly and dislike light. To prevent them causing damage, keep cupboards, shelves and books dry and well aired. Vacuum thoroughly, especially around book shelves and in cracks and corners. After this, give a spray of eucalyptus oil. Silverfish also dislike Epsom salts, cloves, lavender and citronella oils.

To kill, sprinkle borax or alum powder behind books and in corners and crevices. The silverfish will lick the poisonous dust off their feet.

Or, soak rice paper in an alum solution, then dry and place behind books. The silverfish will eat the rice paper and the alum will kill them.

Or, mix together 4 parts borax, 3 parts flour and 1 part sugar. Place the bait in small lids in cupboards, drawers and shelves out of reach of young children and pets.

Place books that have been infested with silverfish in direct sunshine for 3 hours to kill eggs and larvae.

Slaters

Slaters have grey, oval bodies and are able to roll themselves into a ball when disturbed. They are attracted to moisture, so seal entrance points and keep surfaces as dry as possible.

Deter slaters with salt or eucalyptus oil.

Mix up a bait for them of 1 part natural pyrethrum to 2 parts flour. If the position is draughty, mix a wet bait; otherwise it can be dry. Keep out of reach of young children and pets.

Spiders

Learn to live with a few spiders as they help to control silverfish, cockroaches, moths, flies and mosquitoes.

Reduce the breeding places of spiders by cleaning up dark moist places beneath rubbish and in and around garden sheds. Wear gloves when gardening or handling wood or rubbish.

Check gum boots and coats for spiders, especially if they have not been used for some time. Any clothes or bedding left on the floor (particularly overnight) needs to be shaken and inspected carefully.

Remove or kill spiders inside using a broom, fly-swat or vacuum cleaner. Or, trap them using a glass and a piece of paper, then relocate outside.

The green way to spray potentially harmful spiders (such as red-backs) is to use a simple soap spray. It's amazingly effective! See page 123 for the recipe.

Termites

Termites are also known as white ants, although they are not true ants. It is widely known that these social insects, living in their large complex colonies, cause severe damage to buildings. However, it is less well-known that there are alternatives to the use of toxic sprays. A physical barrier is all that is required. Three green products spring to mind, all permanent and all safe and effective:

- *Termi-Mesh* is fine stainless steel mesh (marine grade) that termites cannot penetrate, eat or destroy. It has been proven 100 per cent effective, is fully accredited, conforms to all the necessary codes and can be used for any type of structure. It lasts the life of your home and is installed by accredited tradesmen.
- *Granitgard (granite)* and *Termagard (basalt)* are carefully crushed and graded granite or basalt that are laid as a continuous barrier below the building, including footings, stumps and service pipes. They can be used in the construction of new buildings or when restumping an old home. Both are fully accredited. The particular size of the particles is crucial. They need to be too big for the termites to move, and too closely packed to allow passage between the grains. And the product itself is too hard for termites to chew into. What could be safer? What could be more effective?

Other preventative measures include removing unwanted timber, dead trees and roots from around the house; avoiding landscaping right up to the house slab; and a thorough annual inspection of all timber in and around your home. Raised mud tunnels running along timber framing and up stumps indicate termite activity. Borers, on the other hand, line their tunnels with fine sawdust.

As termites prefer warm, dark places that are damp, take care that timber structures are well ventilated and dry. Firewood should be stacked away from the house and the pile rearranged every 6 months. Any light, hollow-sounding wood should be viewed with suspicion. Avoid storing anything made of wood beneath the house and install ant-caps to stumps in order to help identify a termite problem in its early stages.

White cypress pine and ironbark are relatively resistant to termite attack.

Methods of solving an existing termite problem include using a reputable and registered pest control operator. You may instruct to have the nest located and destroyed rather than the whole house sprayed. A nest can often be dug up or knocked down, then the termite colony killed using boiling water to which some detergent has been added. Alternatively, the broken-up nest can be 'drowned' in a large container of water or sprayed with pyrethrum.

Did you know that termite nests can be sniffed out using specially trained termite dogs?

Wasps

European wasps (with their distinctive yellow and black markings) are aggressive and are attracted to backyard barbecues where meat, salads, sweet foods and drinks are often in abundance. An open drink can or bottle is potentially dangerous, for a bite in the throat can be life-threatening. Always use a drinking straw when consuming drinks outdoors – it's too narrow for a European wasp to pass through.

European wasp nests are best removed at night, and preferably during winter when the colony is inactive. Follow these steps to remove a troublesome nest:

- Cover a torch with red cellophane, as red light doesn't attract these wasps.
- Remember to wear protective clothing and, if stung, apply an ice-pack. Unlike honey bees, wasps are able to sting repeatedly.
- Spray the nest site until it's saturated, using natural pyrethrum.
- Now remove the nest.
- Finally, kill the colony using boiling water to which you've added some detergent. Alternatively, 'drown' the broken-up nest in a large container of water or spray it thoroughly with pyrethrum.

If you're not confident about this process, contact your local council or pest control operator.

Most other wasps are docile and are the natural predators of spiders, flies, mosquitoes and many other insect pests. To remove or kill wasps indoors, use a broom, fly-swat or vacuum cleaner.

Where a nest is in a wall cavity, a simple cardboard cone can be placed over the entrance to the nest. Seal the large end of the cone to the wall material using adhesive tape. Position the small end (1 cm diameter) so that it faces outwards. Now the wasps (or bees) can leave the nest but not re-enter. Gradually the nest becomes unoccupied.

Weevils

It's the larval stage of this beetle that causes the damage. Legless, fleshy, pale-bodied grubs deposit minute droppings and web-like secretions throughout the food on which they feed.

Place bay leaves in your pantry to repel weevils. Or, hang small cotton bags of black pepper in food cupboards.

If your foodstuffs still show a 'webbed' appearance, change your supplier, give your pantry a thorough clean and ensure that all food is stored in sealed containers with bay leaves. Further weevil attack should cease. It is unsafe to use food products contaminated with weevils: their droppings are toxic.

Use your freezer to prevent contamination of grain products. Simply place the grains, flours and cereal products in your freezer for 48 hours, then keep refrigerated until used. No eggs or larvae will survive this chilling process. Alternatively, microwave food for 2 minutes, then store in your refrigerator until it is time to eat it.

Robin's favourite green hints

It wasn't until I began researching this book that I realised how many of the routines practised by women of my grandmother's era were in fact based on sound pest control principles. For in those days, harsh modern cleaners and pesticides had not been invented.

Until I learned that ironing kills the egg and larval stages of many troublesome pests (house dust mite etc.), ironing pillow cases had seemed a case of making work for myself.

Likewise, hanging things out in the sun. The egg and larval stages of fleas, clothes moths, carpet beetles and house dust mites cannot tolerate direct sunlight.

We haven't used the usual flea products on our dogs for over 10 years and we've managed to keep them flea-free. All we use are eucalyptus oil and natural pyrethrum, as well as paying conscientious attention to our dog and cat's indoor and outdoor environments, their bedding, grooming and bathing. Vacuuming and the sunshine treatment have also been important factors. When we lived on King Island, our dog kennels were tucked amongst wormwood hedges, tea tree, cypress and beds of peppermint. Our dogs had no fleas at all, in spite of dry sandy conditions in and around the kennels. Frequent salt-water swimming may have helped as well. It can be done!

Our Siamese cats live indoors with free access to an enclosed garden area via a cat flap. This ensures that both the birds and cats are safe, and that rats and mice stay well away.

To keep ants away from cat food, the feeding bowl is placed in a saucer of water. To repel problem ants, eucalyptus oil is sprayed along their trail.

When we lived in a mosquito area, we used a net over our bed to ensure a good night's sleep. As we rested, the mosquitoes threw temper tantrums outside the net!

Muslin bags containing cloves hang in all our wardrobes to discourage moths. The smell is pleasant and reminds me of the apple pies of my childhood, baked by my aunt.

The smell of mint, rosemary and thyme, when I crush their leaves between my fingers, is a natural and very pleasant way to keep mosquitoes and flies away.

Chapter 9
Your Garden

For most of us, gardens form an important part of our lives. Within your family oasis, you have the ideal opportunity to work with nature to create a garden environment where plants and trees thrive without pesticides.

Sometimes it is difficult to be organic, to care properly for our environment. Often it involves more physical work; always it requires more intellect. But the wise use of mulching, the careful selection of ground cover plants and hand-weeding have to be a safer choice than a weed wand.

CREATING A GREEN HAVEN

Strong vigorous plants are less susceptible to pests and disease. Helpful strategies include crop rotation, sound watering and pruning practices, and the wise use of compost, mulch, organic fertilisers, earthworms and green manure crops. This way you'll create a diverse ecosystem where pests are controlled largely by natural predators.

Birds will be attracted to your garden if you provide them with a birdbath. Flowering trees and shrubs will also attract birds, which will feed on nectar, berries and an unbelievably large number of insect pests.

Frogs, ants, beetles, ladybirds, bees, praying mantids, spiders, wasps and lizards – all these and more are the predators of common garden pests and help to maintain ecological balance in your garden. So it's okay to have a few pests around to feed these beneficial garden creatures. Free range ducks, geese and hens can also assist in the management of garden pests.

Intersperse your garden with strong-smelling flowers and herbs: marigolds, rosemary, onions, garlic, lavender, chives, peppermint. Basil is particularly good planted between rows of tomatoes. These 'companion' plants will help repel insect pests.

Plant water-efficient gardens, making good use of indigenous species. If a plant is really struggling and staggers from one problem to the next, it's often a good idea to pull it out and begin again.

Keep your garden clear of rank grass, piles of rotting timber and sheets of iron, thereby reducing the habitat for garden pests.

Before you reach for that spray or bait (organic or otherwise), at least check to see if the creature is harmful. It may even be beneficial! It seems our society feels the need to eradicate anything that creeps, crawls, runs or flies. Perhaps we should learn to live in harmony with our environment. Your garden is a good place to begin connecting more with the natural world. When you restore health to your garden, you restore health to yourself as well.

If garden pests really are 'bugging you', management strategies involve the sensible use of barriers, traps and baits. Only use an organic spray as a last resort.

ORGANIC PEST CONTROL

Pests need to be controlled rather than eradicated. The following hints show you how to deal with garden pests the green way.

Aphids

With a body length of only 1–2 mm, these soft-bodied, sap-feeding insects usually occur in clusters. Ants frequently nurture aphids and scale insects, feeding on their waste: sweet, sticky honeydew. This honeydew waste also feeds sooty mould.

If the infestation is slight, you can squash the aphids between your thumb and finger.

Or, use a fine, high-pressure jet of water to hose them off plants.

Or, use reflective mulch (aluminium foil) to deter them.

Or, 'paint' aphids with methylated spirits. A simple soap spray is also effective, as are eucalyptus oil and vegetable oil sprays (but only use these two during cool weather or leaves may burn).

To prevent aphid infestation, barriers made of non-drying organic glues or grease bands can be used to protect trees and other ornamental plants, such as roses, from ants – and so too from aphids and sooty mould.

Alternatively, plant onions, garlic or nasturtiums beneath plants prone to aphid attack. They won't like the odour released by these plants.

Or, consider buying aphid-eating insects from

your garden centre. Did you know that ladybirds eat aphids? Attract as many aphid-eating insects and birds as you can and their appetites will work to your benefit.

Bats

Flying foxes can be a problem when they destroy the foliage and fruits of particularly valuable heritage trees. Repellent sprays that cover the leaves and branches of trees used as roosts can be effective and include:

- Chilli spray. This has been used in botanic gardens to deter flying foxes. An environment-friendly foliage fertiliser is applied with an added component of chilli powder. Apparently flying foxes dislike the smell and taste and don't like getting the waxy substance on their feet.
- Wormwood spray.
- Quassia spray.

Wormwood and quassia taste and smell extremely bitter, and everyone knows what happens when you sample chilli. None of these will harm the flying foxes – each will simply put them off eating the leaves. See pages 122 and 124 for quassia and wormwood recipes. Sprays need to be applied when the flying foxes are not roosting.

For orchardists, the only real solution is bat-proof netting. This will enclose the orchard in a protective tent that will also shield the fruit against bird damage, wind and hail. Strange noises and flashing lights only fool flying foxes for a few days – they are too intelligent to be tricked in this way.

Insect-eating bats can be a nuisance if they take up residence in your roof cavity. To solve this problem:

- Hang material such as shade cloth or hessian in front of the entrance points, making the material larger than the hole. The bats can then push out under the fabric but can't find their way back in again. Physically removing the bats will stress both them and you, whereas this method encourages them to search for an alternative roosting site.
- When you are certain that all the bats are out, remove the fabric and block up the holes using aluminium foil or something more substantial.

An absence of tree hollows forces insect-eating bats to roost in house ceilings and other unsuitable places. By placing bat boxes (or hollow logs) in trees or hedges in your garden, you'll be encouraging insect-eating bats to play their part in controlling insect pests.

Note: Make sure that it isn't the breeding season when you follow these hints; otherwise you may cause young bats to be left to starve.

Birds

Birds should be encouraged to live in your garden, as they eat many garden pests as well as provide beauty and song. Make them welcome by creating several sources of permanent water in the form of ponds or birdbaths in positions that are safe from dogs and cats. Did you know that birds attack fruit when their water supply is limited?

You may, however, find it necessary to use deterrents to protect fruit trees at critical times. These need to be changed regularly, as birds become accustomed to their presence or noise. Deterrents include nets; lengths of fishing line, hung loosely so that birds do not get hurt; scarecrows; humming tape; transistor radios; strips of aluminium foil; and tin cans dangling from branches. These can all be effective for a limited time. Electronic, animated scarecrows are another avenue to explore, along with life-sized hawk bird scarers.

To prevent blackbirds digging up your garden, place chicken wire over the surface with holes cut out for your plants. It's worth remembering, though, that all birds, including blackbirds, eat enormous numbers of insect pests.

To prevent bird droppings on car duco, attach loose netting to your carport ceiling.

Cane toads

Cane toads in your garden can pose a problem. The following steps offer a non-violent, humane way to dispose of cane toads:

- Pick up the cane toads in plastic bags used as gloves.
- Seal the bags, then place them in your freezer. (For large-scale collections, plan to have a freezer empty and ready to receive the toads.)
- After a few days, bury the dead toads.

Never allow children to tease or be cruel to cane toads.

If you find long strands of cane toad eggs in moist, watery places and are certain that you have identified them correctly, use a grass rake to pull them free of the water and let them dry out in the sun to die. This is a humane method of killing huge numbers of cane toads.

The call of the cane toad is an unmistakable 'pop-pop-pop'.

Caterpillars

Hand-pick caterpillars off your plants and squash them underfoot.

Or, lightly dust them with flour.

Or, use a weak clay or hot water spray.

Or, use Dipel.

Young seedlings can be protected from caterpillars with mosquito netting.

To deter cabbage white butterflies and cabbage moths (which lay the eggs that produce the caterpillars), make fake life-sized cabbage white butterflies to place amongst your cabbage, broccoli and cauliflower plants. Cut the fake butterflies from aluminium drink cans, paint them creamy-white with black spots and attach them to sticks poked in the ground using fine fishing line.

Or, place nylon pantyhose over newly formed heads. The nylon will stretch as the cabbage grows.

Or, tie a large rhubarb leaf over the top of each newly formed head.

Or, use reflective mulch (aluminium foil) to confuse the butterflies and moths.

Codling moths

The eggs laid by this moth hatch into caterpillars that burrow into fruit, especially apples, although they're also attracted to pears, quinces, peaches, plums and nectarines. When mature (and after having ruined the fruit!), the caterpillars either drop to the ground or crawl down the trunk looking for a place to spin their cocoon. To break the life cycle – by trapping caterpillars on the trunk seeking a sheltered place in which to pupate –

use one of the following trunk wraps:

- Corrugated cardboard bands. Wrap long strips around the trunk, with the exposed ridges facing inwards. Tie firmly with string.
- Newspaper, several layers thick, tied around the trunk of the tree.
- A hessian sack or strips of hessian wrapped around the trunk.

Remove wraps every 4–6 weeks and squash any cocoons you find. Now renew the trunk wraps to trap the next generation. Three hatchings are possible during the one season. Begin using the trunk wraps immediately after flowering.

Other green strategies that break the life cycle include:

- Place a mulch of fresh green lawn clippings beneath fruit trees, especially during summer. The hot compost effect will kill any mature caterpillars that crawl beneath it to pupate. Hot compost decomposes organic matter faster than normal due to the presence of micro-organisms living in ideal conditions: moisture, warmth, air and nitrogen – in this case provided by fresh grass clippings, although it could also be supplied by blood and bone or animal manure. Hot compost will also destroy disease organisms and weed seeds.
- Prevent your trees from fruiting for one to two years.
- Use simple pheromone traps to lure males to their death (by adhesion to a sticky pad), thereby reducing the likelihood of females laying fertile eggs.

Some home gardeners cover their fruit trees with insect-proof netting, while others surround individual branches of fruit with fine-mesh bags. The advantage of these methods is that they also prevent attack by birds, European wasps and fruit fly. These measures need to be put in place early in the flowering stage, as the female moth lays her eggs inside the developing flower.

Every day, pick up and destroy all fallen and infested fruit by boiling, feeding to farm animals, immersing in water for at least 3 days, or sealing in a plastic bag and leaving in the sun until the fruit goes rotten. This will prevent mature caterpillars entering the soil to pupate and so break their life cycle.

Earwigs

Earwigs run quickly and are easily recognised by the pair of large curved pincers on the end of their body. To deal with plague numbers, make a trap using either:

- Pieces of thin black polythene pipe (or short pieces of old garden hose) closed at one end. The earwigs will shelter here during the day.
- Empty flower pots with the drainage holes sealed. Fill them with dry grass or crumpled newspaper and place upside down where earwigs are a problem. To allow the earwigs entry, prop them up a little using sticks or small stones.

Empty both types of trap every few days, and destroy contents.

Or, spray with eucalyptus oil or soap spray to kill.

If earwigs are nibbling rosebuds, place a sticky or greasy 'collar' around the base of the main stem of each bush.

Fruit fly

These flies are strongly attracted to sweet, yeasty mixtures, so choose any of the following liquids to lure them to death by drowning:

- Bran and sugar dissolved in hot water.
- Flour, molasses and water.
- Vegemite, banana peel and water.

Early in the season, it's a good idea to thin the fruit on your trees and then, if possible, place fine-mesh or paper bags securely over the remaining fruit to exclude fruit flies and birds. Another option is insect-proof netting over trees.

Destroy all fallen and infested fruit by boiling, feeding to farm animals, immersing in water for at least 3 days, or sealing in a plastic bag and leaving in the sun until the fruit goes rotten. This will stop fruit fly maggots entering the soil and continuing their life cycle.

Grasshoppers

Hedges act as natural barriers against many pests. Use plants with tough foliage and strong fragrances, such as rosemary, wormwood and lavender.

Free range hens scratching around a fenced-in vegetable garden will help protect the plants. Keep the hens moderately hungry and feed grain as close to the vegetable patch as possible.

Guinea fowl or turkeys pursue grasshoppers with even more enthusiasm than hens! Ducks and geese also assist in pest control.

A fine mist of water will help protect gardens from grasshoppers, as these insects dislike flying through water.

Shadecloth or fly-netting can be used to cover particularly valuable plants.

Harlequin bugs

Trap these colourful, sap-sucking bugs (often found in clusters) beneath pieces of cardboard placed on the ground. Check the traps in the late afternoon and either stamp on the bugs or drown them.

Or, hose them off your plants, then stamp on them. Be sure to look under the leaves.

As a last resort, use eucalyptus oil spray or pyrethrum spray.

Mites and other tiny insects

Simply blast them off foliage and stems with a fine, high-pressure jet of water. Mites don't like getting wet. Use an old toothbrush to remove those clinging too tightly to be hosed off.

Or, use a soap, milk, clay or pyrethrum spray.

Pear slugs (larvae of sawfly wasp)

Pear slugs are also found on cherry, plum, quince, apple, almond, crab apple, hawthorn and rowan trees. These small flat slugs are covered with a shiny, dark, green-brown slime. Clinging to the upper surfaces of leaves, they feed on the green surface tissue until only a fine network of veins remain.

To prevent defoliation of trees, dust leaves with fine wood ash. Its gritty texture is lethal to the slug.

Alternatively, use a mixture of fine sand and lime, or lightly browned flour. Dust regularly as the dust will not kill the eggs.

Or, blast the slugs off the leaves using a jet of water from your hose.

Free range hens, as well as other birds in your garden, enjoy eating these soft slugs.

Possums (and rabbits)

Possums can be a nuisance if they take up residence in your roof. There are, however, solutions that are both humane and practical.

Firstly, organise alternate accommodation such as possum boxes in trees or hedges. Ensure these are safe, cosy and waterproof. You can purchase boxes or make them yourself.

Secondly, identify the main exit point (the one with tufts of fur around it) and block off any other openings under the roof with netting or timber. After 1 week, and between 10 p.m. and midnight, close off the main exit point.

Thirdly, check that you've removed any overhanging branches or creepers that give possums access to your roof.

Trunk collars and barriers
Use trunk collars, made of heavy, clear plastic sheeting or aluminium or sheet iron, to protect fruit and ornamental trees from possums. These should also be used around the trunks of trees that give access to your roof. Collars need to be 60 cm wide and placed at least 1 metre above the ground.

Rabbit net your boundary fence (burying the netting a third of a metre into the earth) to prevent rabbits entering.

Repellents
Try spraying favourite plants with a white pepper and soapy water solution.

Or, make up a quassia chip solution to spray around the edge of your garden or on target plants. Possums, rabbits and birds dislike the bitter taste. Alternatively, use a wormwood spray or dust. Like quassia, it's very bitter tasting and smelling.

Or, sprinkle blood-and-bone mixture around plants and over the foliage of roses and seedlings (vegetable and flower) to deter both possums and rabbits. Repeat every 2–3 weeks if rain washes it off. Alternatively, fill pantyhose with blood and bone, then hang in trees or treasured shrubs.

If you regularly feed resident possums, their appetite for roses and precious shrubs will decrease. Try sweet treats such as bread and jam (especially apricot!), fruit cake or any ripe fruit. These delightful marsupials will reward you with their friendship.

Remember that possums are very territorial, so evicting them from your property is futile. Another pair will move in the moment the territory is vacant!

Scale insects

These tiny sap-sucking insects are usually found clustered on the undersides of leaves. They secrete sweet, sticky honeydew as waste, which attracts and feeds both ants and sooty mould. There are several ways to deal with this problem.

If the infestation is slight, squash scale insects between your fingers.

Or, scrub insects carefully from twigs and leaves using a firm brush and warm soapy water.

Or, spray with soap or clay spray to suffocate. Vegetable oil and eucalyptus oil sprays are effective as well, but use only in cool weather to avoid burning the leaves .

Or, apply white oil (half-strength) to kill. This has a relatively short residual life.

Cut away badly affected foliage and consider introducing beneficial insects to your garden such as tiny parasitic wasps and ladybird beetles and their larvae. These can be bought from your garden centre and will feast on scale insects, aphids and many other plant pests.

Snails and slugs

Encourage birds, lizards and frogs to live in your garden. They thrive on a diet of slugs and snails, especially the eggs and young ones.

Maybe your garden would benefit from a resident duck? As well as eating many garden pests, a duck can be an engaging pet and will even lay eggs for the table.

Clear away debris and weed growth, as these make ideal breeding habitats for slugs and snails.

Go outside after dark with a torch, preferably after light rain. Stamp on the snails with your gumboots or drown them in a bucket of warm soapy water. Through this method alone, significant numbers of snails can be killed.

A green reminder: commercial snail bait frequently causes the death of pet dogs. If a lizard eats snails poisoned with snail bait, the residual poison kills the lizard as well. Lizards are one of the very best forms of natural pest control.

Traps

Make traps that drown these pests by selecting any wide, shallow and slippery container and sinking it level with the ground. Fill with a mixture of 1 part water, 1 part beer (it can be flat) and some brown sugar. You can safely feed your catch to birds.

Barriers
Create protective circles of soot, lime, sawdust, grit or wood ash to protect seedlings and delicate shrubs. The grit will stick to the slimy surface of slugs or snails and deter them. Alternatively, circle seedlings with cat litter (white clay particle type). The extremely absorbent nature of the litter deters slugs and snails as they need moist surfaces on which to move.

Or, place wormwood prunings around seedlings. Slugs and snails are repelled by the strong odour of wormwood.

Or, cut strips of fly-wire (not plastic or fibreglass), then roll each into a circle and join. Pull two or three wires off one end of your fly-wire tubes then place these snail 'tree-guards' over young lettuce plants. The longer vertical wires are placed on the top to prevent snails crawling over.

Or, use cut-off plastic drink bottles to protect young seedlings. These act as mini-greenhouses as well.

HOMEMADE ORGANIC SPRAYS

Organic sprays break down more rapidly than commercially produced pesticides and can therefore be labelled 'environment-friendly'. Although these sprays are made with 'natural' ingredients, they may, however, be toxic both to humans and to garden-friendly creatures – as well as to the organisms you seek to control. Remember, too, that pests develop resistance to sprays after repeated exposure to them.

You should always wear protective clothing, a face mask and gloves when applying sprays such as quassia, pyrethrum, wormwood and Dipel. Also, label and store organic sprays as carefully as you would other commercially produced pesticides and keep them well away from children and pets.

The philosophy, 'If a little spray is good, then a lot must be better,' is not a wise one, regardless of whether the spray is a commercial preparation or a homemade organic one.

Before you reach for a spray, ask yourself the following questions:

- Is it possible to remove the pests by hand, by shaking, by vacuuming, by a strong spray of water or by squashing them? For example, have you tried a night collection of pests such as slugs and snails, especially after rain?
- Can you prune away the diseased or insect-damaged leaves or branches?
- Have you tried barriers to prevent the pest reaching the crop? For instance, fine-mesh, insect-proof netting in fruit trees; protective circles of lime to deter slugs and snails; stainless steel mesh to protect house foundations from termite attack.
- Have you tried traps? For example, stale beer (equal parts beer and water, with some brown sugar) to lure slugs and snails to their death.
- Would a tree trunk band or a sticky, glue-covered band enable you to catch and destroy the nuisance species, so breaking its life cycle? For instance, a tree trunk band of corrugated cardboard will catch the codling moth in its cocoon stage.
- Would soil solarisation destroy nuisance weed

seeds; or sunshine kill house dust mites living in your rugs?
- Have you made full use of repellents such as eucalyptus and wormwood for fleas; cloves for clothes moths; and blood and bone for possums?
- Would either an early or late planting help miss the main flush of, for example, the cabbage white butterfly?
- Have you controlled weed growth (hand weeding is effective!) and removed fallen fruit, diseased leaves and plant waste?
- Have you made use of sheets of aluminium foil or blue plastic, spread between rows, to reflect light and confuse insects such as aphids?
- Did you know that aphids and whiteflies (both sap-suckers) find the colour yellow irresistible? Simply paint some timber yellow, then thickly coat it with vegetable oil and you'll trap lots of these pests! There are also commercially produced sticky yellow traps. They will attract and trap European wasps as well.

Sprays seem an easy answer but should be used only as a last resort – and then sparingly.

Recipes for some relatively low-impact organic sprays are given below. When water is required for a recipe, soft water (preferably rainwater) is recommended.

Baking soda (bicarb soda) spray

This spray prevents fungal spores establishing themselves and developing on your plants. It is especially effective in treating any fungal or mildew problem on grape or passionfruit vines.

Simply combine 1 teaspoon of bicarb soda with a few drops of liquid soap, then dissolve in 2 litres of water. The soap helps the spray stick to the leaf surface. You can spray twice-weekly with this solution.

Chamomile spray

This easy-to-make spray acts against powdery mildew, rust, stem rot, brown spot, brown rot, leaf spot and other fungal diseases. It is the gentlest fungicide possible.

Simply make up a pot of ordinary chamomile tea, then leave it to brew for 10 minutes. Cool, then spray every few days.

Clay spray

This spray suffocates creatures such as mites, thrips, caterpillars and aphids; however, remember that useful creatures such as ladybird larvae will be affected as well, so restrict your spraying to creatures you can actually identify. The spray has no residual effect, so it can be re-applied every few days.

Using pure clay, dilute the clay with sufficient water to make a spray. It's as simple as that.

Diatomaceous earth spray

This fine dust is a non-toxic product mined from the fossilised remains of an algae known as diatom – so it's actually the ground-up skeletons of marine organisms. With its microscopic, sharp edges, the fine powder pierces cuts into any soft-bodied insect, eventually causing death from dehydration.

This product, used as a dust or spray, is effective against caterpillars, pear and cherry slugs, slugs, snails, termites, aphids, mites, thrips, silverfish and even cockroaches.

Due to the irritant nature of this dust, it's very important to protect your nose, eyes and mouth from exposure, especially when making up the spray or slurry or when using the material as a dust. Apply the dust after light rain, so that it sticks well.

To make up your spray, mix 30 grams of diatomaceous earth with a quarter-teaspoon of liquid soap. Now add 4 1/2 litres of water and mix well. The soap helps the spray stick to the insect's body.

Another way of using this material is to paint a thick slurry around shrub and tree trunks (as a collar) to protect them from caterpillar attack.

Dipel spray

This is a commercial product containing spores of the bacterium *Bacillus thuringienis*. It's an excellent example of biological control, being largely non-toxic to mammals. Lethal effects are not passed on down the food chain to birds or any other predator.

Ideally, Dipel spray needs to be re-applied about every 7 days, as its spores are destroyed by sunlight. It is especially effective against moth and butterfly caterpillars, which readily ingest it, resulting in paralysis of their digestive tract.

Eucalyptus oil spray

Eucalyptus oil, like many other essential oils, kills scale insects, aphids, earwigs, slugs, slaters, whiteflies, mites and many other pests. It is a safe, non-residual spray, best applied around seedlings and at the base of plants.

To make up the spray, combine 1 teaspoon of eucalyptus oil with 500 ml of soapy water.

Generally speaking, a solution of about 2 per cent eucalyptus oil in water is considered a good general purpose insect spray. You can repeat the spray every 3 days.

Oil sprays work better in the winter than the summer months, when temperatures from the mid-20s upwards may cause the oil to burn the leaves. Another reason for applying this spray in winter rather than summer is that leaves are more porous during the cooler months. The spray breaks down quickly.

Hot water spray

Many soft-bodied insects are killed by a simple spray of hot water (45° C to 55° C). This will not harm most foliage.

Milk spray

Milk is lethal to red spider mites and mildew. A milk spray can be used on plants such as zucchinis, lettuces, cucumbers and tomatoes.

To prepare your milk spray, simply mix 1 part milk to 10 parts water. The spray needs to be repeated every 2 weeks.

Pyrethrum (natural plant-derived) spray

Made from the dried, finely ground flowers of pyrethrum daisies (*Chrysanthemum cinerarifolium* and *C. coccineum*), pyrethrum is an insecticidal dust that attacks an insect's central nervous system. It kills flies, mosquitoes and other chewing and sucking insect pests. It breaks down rapidly in heat and light.

As pyrethrum is moderately toxic to mammals and highly toxic to fish and frogs, use it only as a last resort – and avoid using it near ponds and waterways. To make up the spray, follow the directions provided with the product. Commercial dusts and sprays may contain soap, which helps the pyrethrum adhere better to the insect's body.

Spray in the early evening, as pyrethrum is fatal to bees. By the following morning, it will have broken down. This type of spray is rarely needed in the home garden.

Quassia chip spray

Quassia chips come from the bark of a small South American tree, *Picrasma quassioides*. Being relatively safe, quassia has definitely come back into fashion as a repellent. Due to its extremely bitter taste, it is a very effective deterrent to possums, rabbits, flying foxes and birds.

Using an enamel or stainless steel saucepan, boil 50 grams of quassia chips in 4 cups of water for about 1 hour, with the lid on. Cool and strain. Add enough soap to the concentrate so that it lathers when whipped. Use 1 part concentrate to 3 parts water to make a spray. Store out of reach of children and take care not to breathe the fumes while your brew is boiling.

Seaweed spray

Seaweed spray helps plants resist conditions such as curly leaf, brown rot, black spot, powdery mildew and many other fungal and bacterial conditions. It also helps protect against frost damage and provides fertiliser to the plant as well as increasing soil micro-organism activity. It is best used at monthly intervals and sprayed at night during the summer months.

To convert half a bucket of seaweed into a useful seaweed spray, rinse away the surface salt, then cover with tap water. Allow to soak for 3 weeks, then drain off the liquid. Dilute with more tap water until the liquid is the colour of weak tea. It's as simple as that, and you can use the leftover seaweed as an excellent mulch.

Soap spray

A soap spray will kill caterpillars, thrips, scale insects, mites, whiteflies and aphids. It does this by paralysing its victim, which eventually dies of starvation. By killing aphids and scale insects, soap spray also controls sooty mould.

For this spray, mix together soap and water until you have a frothy, milky solution. Allow the spray to dry on the leaves, then rinse the leaves clean the following day. Spray every 2–3 days for two weeks. If your plants are drought- or heat-stressed, or weakened in any way, use a more dilute mixture.

Vegetable oil spray

This spray kills by lightly coating the insect (and its eggs and immature stages) with vegetable oil, with the effect of suffocating it: the oil blocks the insect's supply of oxygen. Aphids, spider mites, scale insects, mealybugs and some caterpillars are affected by this spray.

Oil sprays work better in the winter than the summer months, when temperatures from the mid-20s upwards may cause the oil to burn the leaves. Another reason for applying this spray in winter rather than summer is that leaves are more porous during the cooler months. The spray breaks down quickly.

This oil acts as a barrier to infection and helps prevent fungal rusts and mildews.

Mix 1 cup of vegetable oil with 1 tablespoon of liquid soap. For every 2 1/2 teaspoons of this mixture, add 1 cup of water. This makes a spray that spreads well over most surfaces. The soap helps the spray stick to the insect.

Wormwood spray

This soft silvery-grey shrub (*Artemisia* family) has extremely aromatic foliage that is very bitter tasting. Wormwood is even mentioned in the Bible as a bitter herb. It's easy to strike as a cutting and grows readily in most areas. You have probably seen it at the seaside or around old homesteads, especially in the country.

Wormwood makes an excellent insecticide, killing and repelling insects, especially fleas. It repels birds, mice, rabbits, flying foxes and possums.

To make wormwood dust, simply pick an armful of the plant, then dry in bunches by hanging it in a cool, airy, dry place. After about 10 days, powder the leaves by rubbing them through a fine wire sieve. The fine, very bitter dust can now be used to protect precious plants in your garden. To help it adhere to leaves, apply after light rain or a light watering.

To make a bitter and very effective spray, simply cover your chopped-up leaves with boiling water and leave for 3–4 hours to infuse. Now strain the solution and dilute with 1 part of spray to 4 parts of water.

Summing up

For insects: use clay, diatomaceous earth, Dipel, eucalyptus oil, hot water, milk, soap, vegetable oil or pyrethrum.

For mould and mildew: use bicarb soda, chamomile, milk or seaweed. Bordeaux (a special mix of copper sulphate (bluestone), calcium hydroxide ('Limil') and water) can also be used.

To deter possums, rabbits, birds and flying foxes: use wormwood (which is also excellent for fleas) or quassia.

Expensive commercial pesticides, herbicides, soil sterilants and fungicides pose an unnecessary danger to your family and the environment. There is always a green alternative to using these products.

GREYWATER

Greywater is the term used to describe the waste water from your kitchen, bathroom and laundry. Untreated greywater can be diverted from the sewage system to your garden, especially during dry months when water shortages and restrictions are in place. At all other times, however, this waste water must be directed into the sewage system.

Waste water diversion valves can be purchased from hardware stores. These devices divert greywater away from the sewage system, ideally into slotted plastic pipes (buried beneath mulch) that water your plants. If you lack confidence in this area, a plumber will be able to install a waste water diversion valve for you. There are, of course, other less complex ways of transferring waste water from within the home to your garden. The simplest way is by using a bucket. Another approach is to place a flexible hose between, for example, your bath and garden, then siphon out the water. There is no single greywater system as such.

Untreated greywater from your home must exclude water from your toilet, and laundry water contaminated with faeces from nappies. It is also not a good idea to recycle water from your kitchen, as food particles attract flies, rats and mice, and kitchen waste water also contains other contaminants not readily broken down by soil organisms.

Remember that greywater will contain whatever soaps, detergents, disinfectants and cleaners you use within the home. The use of biodegradable products is, therefore, of special importance. Choose those labelled low phosphate or phosphate-free, and use as little as possible, especially in your washing machine.

Untreated greywater may contain harmful bacteria, viruses and pollutants. These pose health risks (to humans, pets and other animals) and environmental risks. You should therefore check with your council and local water authority prior to diverting any untreated domestic greywater for re-use in your garden.

The following hints will guide you towards the safe re-use of untreated greywater:

- Do not use greywater to grow vegetables eaten either raw or undercooked. You'll need town water or tank water (rain water collected from roof) for these crops.
- Reduce your use of fertilisers, as some plant nutrients will be supplied in the greywater.
- Never use hot greywater.
- Do not allow any animal to drink greywater.
- Wash your hands after handling greywater.
- Never store for more than 24 hours.
- Don't spray plant foliage with greywater.

Instead, apply it close to the ground, preferably beneath mulch.

- Avoid saturating the soil or creating 'ponds'. Aim for trickle irrigation.
- Never allow greywater to enter a neighbour's property or enter a stormwater drain, river or creek.

Using water wisely and planting indigenous gardens with species selected with their water efficiency in mind makes good sense. Keen gardeners, however, will wish to grow exotic species, along with food gardens. These climatically challenged plants will require water during dry months and periods of drought, and the use of greywater in these circumstances can prove invaluable.

OTHER GREEN OUTDOOR HINTS

Barbecue cooking plate (to clean)

Sprinkle salt (to absorb grease and dirt) on the barbecue plate while it's still very hot, then leave to cool before brushing clean. Protect from rust by applying a thin film of vegetable oil.

Concrete paths

To clean stains from concrete paths, wash with a solution of 5 tablespoons of borax to 5 litres of hot water. Unpleasant smells may be removed by smearing a bicarb soda paste over the stain, letting it dry, then brushing clean.

Grass and weeds in paths and driveways can be controlled by hand-weeding, hoeing or spraying with super-hot water.

Oil or grease, spilled on concrete, sandstone,

terrazzo or bricks, may be removed in the following way:

- Absorb surface grease with paper towelling. Alternatively, sprinkle with cat litter (white clay particle type) or dry sand, then sweep clean.
- Combine equal parts chalk and bicarb soda with enough water to make a sloppy paste.
- Spread paste over the stain. Seal with plastic.
- Leave until completely dry.
- Brush (or vacuum) away the chalk and bicarb soda, which will have absorbed the oil or grease. Repeat if necessary.

You can also use this poultice to remove grease or oil stains from garage floors, and unsightly soot or smoke marks from stone or bricks laid around fireplaces.

Lichen

The greenest way to remove lichen from roof surfaces is by water blasting.

Moss

To encourage the growth of moss, 'water' with milk.

To kill moss (on a path or paving), make up a concentrated salt and vinegar solution by dissolving salt in boiling water, then adding some vinegar. 'Paint' the moss with a brush dipped in the solution. When the moss is dead, sweep clean using a straw broom.

Or, spray moss with 1 part vinegar and 1 part methylated spirits. Repeat in half an hour. Sweep clean when dry.

Powdery mildew (especially on zucchini and cucumber plants)

Natural products such as milk, chamomile, bicarb soda and seaweed can be used as sprays to treat powdery mildew and other fungal problems. Chamomile is the gentlest fungicide possible, but you will need to spray every few days.

Avoid overhead watering, which spreads fungi spores. Rotate crops, use regular seaweed sprays to increase resistance, improve air circulation, mulch well and cut off any infected leaves. Place mildew-affected leaves in hot compost to kill the spores.

Swimming pool tiles

To clean swimming pool tiles, use bicarb soda on a soft cloth.

Treated pine

The treated pine process involves the use of an arsenic compound to preserve the timber. In the presence of water, this compound has been shown to leach from wood. In the presence of fire, dangerous fumes are produced. Therefore, do not burn treated pine or use treated pine for your compost bin, worm farm or vegetable garden beds.

Weeds

Removing weeds from the ground need not rely on weedicides. The following are green solutions:

- Hand-pull all shallow-rooted plants.
- Gently pull back all tap-roots, tubers, corms and runners along the growth line, or dig them up with a mattock or fork, then hang off the ground to dry out in hot sun or a drying wind.

- Whipper-snip repeatedly or mow emerging stems.
- Allow continual browsing by animals such as sheep and goats. Goats do an excellent job with blackberries.
- Cover with carpet or black plastic, or with clear plastic as a form of solarisation. A 1 cm thickness of newspaper used as a weed control mat is also effective, along with deep mulch up to 10 cm in depth.
- Cultivate repeatedly to prevent weed re-growth.
- Cut off plants such as pampas grass at ground level, then burn the stump.
- Killing weeds with super-heated hot water is an excellent method of weed control. A commercially made green weeder is now available that produces super-heated hot water to spot-kill weeds instantly and safely by breaking down plant cells. It will kill weeds between pavers, along fence lines, around posts, around the bases of trees and shrubs, and in lawns. And you don't need to bend or kneel down, or wear a mask. But you do need eye protection.

Remember the age-old rule: never allow a weed to seed.

Worm farm

A worm farm that recycles all your household scraps and organic waste can also provide fertiliser (castings) for your garden. Worm farms are an excellent idea.

Robin's favourite green hints

Our native garden is alive with wrens, honey-eaters and fly-catchers, and a thriving population of blue-tongued lizards makes sure that slugs and snails never nibble the vegetables we grow. To deter the occasional slug or snail from eating sensitive young seedlings, we enclose the seedlings within a boundary of lime or use wormwood prunings to create protective barriers. Our dogs, by their mere presence, keep the rabbits away.

On King Island, grasshoppers were a seasonal problem in our vegetable garden. By allowing our hens to free range around the garden and by feeding them next to the vegetable patch, we prevented the grasshoppers from stripping our garden.

We have always preferred to share a certain percentage of fruit and vegetables with resident possums and birds rather than getting frantic over a few losses. At Longwood, we had a small orchard. By draping the trees with old curtain netting, the parrots left most of the fruit; and by having our dogs free within that area, we kept the possums under reasonable control.

Tomatoes seem irresistible to blackbirds, yet by leaving them a feed and picking the rest when the colour first comes through, we save most of the crop. A birdbath close by ensures that they are not eating tomatoes because they're desperate for moisture.

With the deadline for this book fast approaching, there seems to be no time for anything else. And it's increasingly obvious that there are many, many things around our home and garden that need doing! But that's okay. I don't expect or want to live in a 'perfect' house with a 'perfect' garden. I prefer to live in a *home* that's people-friendly and environmentally safe – with an easy-care garden that's a busy, balanced, blend of birds, lizards and insects.

A sense of connection with the natural world ensures that life is rich and balanced. Living green improves your quality of life – lets you feel whole again.

Products and Where To Find Them

Note: A large number of these products are by no means essential, as many eventualities described in this book will never occur either in your home or garden. It's suggested that you purchase these items only if and when you need to do so.

Garden Centre

- Beneficial insects, such as lacewings, ladybird-type beetles and their larvae etc.; also available through horticultural suppliers
- Bordeaux
- Companion plants such as marigolds, rosemary etc.
- Dipel
- Lime (calcium carbonate)
- 'Limil' (calcium hydroxide)
- Pheromone traps; also available through horticultural suppliers
- Pyrethrum (natural)
- Sticky yellow traps; also available through horticultural suppliers
- Quassia chips

Hardware store

- Alum (potassium alum) powder
- Chalk
- Greywater diversion valve
- 'Limil' (calcium hydroxide)
- Linseed oil
- Neatsfoot oil
- Plaster of Paris

Health Food Shop

- Bay leaves
- Bicarb soda (sodium bicarbonate), in bulk
- Biodegradable dish-washing detergent, laundry powder and liquid, and pet products
- Essential oils (tea tree, citronella, eucalyptus, lavender etc.)
- Herbal teas and dried herbs
- Linseed oil
- Manuka oil
- Neem oil
- Pure soap
- Quassia chips
- Rice paper
- White vinegar, in bulk

Hobby supplies

- Beeswax
- Clay
- Plaster of Paris
- Chalk
- Wood ash

Pharmacy

- Alum (potassium alum)
- Borax
- Epsom salts (magnesium sulphate)
- Essential oils (tea tree, citronella, eucalyptus, lavender etc.)
- Lanolin
- Mineral oil
- Neem oil
- Paraffin oil
- Pure soap

Pools supplies

- Diatomaceous earth

Supermarket

- Bay leaves
- Bicarb soda (sodium bicarbonate), with baking products
- Borax, with laundry products
- Cat litter (white clay particle type)
- Cloves, with herbs and spices
- Epsom salts (magnesium sulphate), with medical products
- Eucalyptus oil, with medical products
- Fly-swat
- Methylated spirits
- Molasses
- Pure soap
- Rice paper
- Washing soda (sodium carbonate, 'Lectric' soda), with laundry products
- White vinegar

Acknowledgements

My husband, Doug, has for the past 36 years shared my passion for all things natural and the environment. Together we have made this journey of discovery – created this book.

A sincere thank-you to my friends and relatives who gave so generously of their time and hints. Also, to the authors of books I've absorbed over the years.

Lastly, I'd like to thank my editor, Chris Feik, for his editorial expertise. Also, the entire team at Black Inc., who are always a pleasure to work with.